ESSENCE of CREATIVITY

ESSENCE
of
CREATIVITY

A Guide to Tackling
Difficult Problems

STEVEN H. KIM
Massachusetts Institute of Technology

New York   Oxford
OXFORD UNIVERSITY PRESS
1990

Oxford University Press

Oxford   New York   Toronto
Delhi   Bombay   Calcutta   Madras   Karachi
Petaling Jaya   Singapore   Hong Kong   Tokyo
Nairobi   Dar es Salaam   Cape Town
Melbourne   Auckland

and associated companies in
Berlin   Ibadan

Published by Oxford University Press, Inc.,
200 Madison Avenue, New York, New York 10016

Oxford is a registered trademark of Oxford University Press

Library of Congress Cataloging-in-Publication Data
Kim, Steven H.
Essence of creativity : a guide to tackling difficult problems /
Steven H. Kim.
p. cm.   Bibliography: p.   Includes index.
ISBN 0-19-506017-2
1. Creative ability in business 2. Decision-making.
3. Industrial project management. 4 Information technology.
5. Technological innovations. I. Title.
HD53.K56   1990
658.4'03—dc20   89-16144   CIP

to my sister Jean

Preface

Like so many things in life, challenging problems are a boon and a bane. Difficult tasks can unleash cyclones of frustration: they accelerate our pulses and sometimes leave us with ulcers, and they will probably curtail our individual life spans. On the other hand, the knotty problems of life offer us food for thought, sustain our creativity, and reaffirm our self-ordained title as *Homo sapiens*. They add emotion and spice to the human experience. To adapt an adage, It's better to have lived and struggled than not to have struggled at all.

When the most obstinate puzzles of personal life and professional work succumb to our machinations, the feeling of exhilaration is compelling. After all is said and done, we realize that the troughs of anguish and despair during the problem-solving phase were—well—not without reward.

In fact, some of us find problem solving so addictive that we while away our days in laboratories, computer rooms, and other niches where stubborn problems abound. Some would sooner trade riches, influence, and respectability for the opportunity to tackle impossible problems. (Others, of course, settle for all of the returns above.) What is money when you can have instant access to unlimited problems?

We encounter knotty problems day in and day out. The solutions to these problems are unclear *a priori;* hence they must be sought with resourcefulness and creativity. Despite the prevalence of such tasks, we have little insight into the nature of difficult problems, and even less into methods for their resolution. This book represents an initial step toward developing a coherent framework for the nature of knotty problems and the dimensions of creative behavior demanded by such tasks.

This work germinated from a set of notes I had prepared to help orient my students and research assistants at the Massachusetts Institute of Technology. As their thesis supervisor, I was expected to guide their multifarious research activities with vision and prudence. It was unclear to me where these qualities were to be procured, so I began to fabricate some guidelines. These took the form of reflections on the research experience, and schemes for tackling impossible questions. The assorted notions were then reduced to ink, as much to clarify my thoughts (see the discussion on output procedures in Chapter 5) as to help or hinder my students.

A special feature of this book is a discussion of research practices whose principles should apply to diverse settings, ranging from the university laboratory to the corporate marketing office and governmental research institute. The topics will be of special relevance to researchers and their supervisors. These concerns, however, also apply to cozier circumstances. Some of the principles have been utilized by students in our laboratory for more constrained applications such as learning a new subject or writing for English class.

A central theme in these pages is the notion that ideas tend to grow incrementally rather than precipitously; and the book itself is no exception. Few, if any, of the ideas in these pages are new. Surely each could trace a lineage to an ancestor at some other place and time. If a creative factor lurks within, a measure of it must lie in the larger organization of the underlying ideas. This book is an attempt to distill the voluminous literature on creativity, innovation and project management into its essence, and to present it as a slim set of working hypotheses and workable practices.

Of special help in writing this book were the advice and reactions of Beth Hennessey. Beth provided high-level structural suggestions as well as detailed criticism of specific passages. Another guiding spirit was Elias Towe, who offered many insights as the scribblings took shape. The manuscript improved in spite of itself as a result of their thoughtful inputs.

I am also grateful to Lindsay Moran, Christina Zwart, and Kristen Svingen for their editorial assistance in preparing the text and figures. Their efforts were supplemented by the contributions of Elisabeth Ellis, Polly DeFrank, and Maureen Kelly.

I gladly take credit for all deficiencies, delusions, and indiscretions of like ilk. Should this slim volume give you cause to ponder or reconsider your problem solving strategies, then my effort in compiling it will have been worthwhile.

Here's to life and its challenges!

Cambridge, Mass.   S.K.
October 1989

Contents

ESSENCE of CREATIVITY

1

Introduction

A thing constructed can only be loved after it has been constructed; but a thing created is loved before it exists.　　　　Gilbert Keith Chesterton

The world around us abounds with problems requiring creative solutions. Some of these are naturally induced, as when an earthquake levels a city or an epidemic decimates a population. Others are products of our own creation, as in the "need" to curb pollution, to develop a theory of intelligence, or to compose works of art. Still others are a combination of both, as in the development of high-yield grains to feed an overpopulated planet, or the maintenance of health in the face of ravaging diseases.

The word *problem* is used in a general sense to refer to any mental activity having some recognizable goal. The goal itself may not be apparent beforehand. Problems may be characterized by three dimensions relating to domain, difficulty, and size. These attributes are depicted in Figure 1.1.

The *domain* refers to the realm of application. These realms may relate to the sciences, technology, arts, or social crafts.

The dimension of *difficulty* pertains to the conceptual challenge involved in identifying an acceptable solution to the problem. A difficult problem, then, is one that admits no obvious solution, nor even a well-defined approach to seeking it.

The *size* denotes the magnitude of work or resources required to develop a solution and implement it. This attribute differs from the notion of difficulty in that it applies to the stage that comes after a solution has been identified. In other words, difficulty refers to the prior burden in defining a problem or identifying a solution, while size describes the amount of work required to implement or realize the solution once it has jelled conceptually.

For convenience in representation on a 2-dimensional page, the domain axis may be compressed into the plane of other attributes. The result is Figure 1.2, which presents sample problems to illustrate the two dimensions of difficulty and size. Cleaning up spilled milk is a trivial problem having numerous simple solutions. In contrast, refacing the subway trains in New York City with a fresh coat of paint is a formidable task that could require hundreds of workyears of effort. Even so, the general approach to sprucing up the subway presents no conceptual difficulties.

When we need to drive up a steep hill covered with ice, it is not clear what the best approach is. Should we start at full speed from the bottom of the hill, and

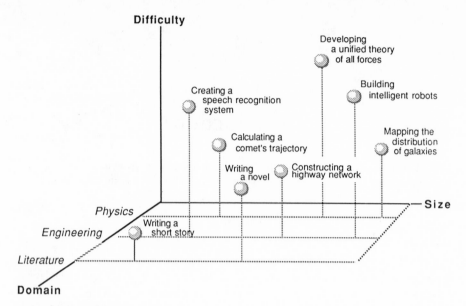

Figure 1.1. Dimensions of problems.

thereby risk running down any incautious pets and pedestrians? Should we attempt to traverse the road in a series of diagonal paths, and risk being hit by a car on the left side of the road? Should we circumnavigate the hill and seek a road with a gentler gradient? Not only is it unclear which is the optimal strategy to pursue, but none of these possibilities offers any guarantee of resolving the problem.

Many problems are both difficult and gargantuan. An example is found in developing a unified theory of intelligent behavior. It is not clear where the solution lies, or even in which direction to best proceed. Further, a complete solution will not likely be found one serendipitous evening, but will gradually evolve over the years through the collective effort of hundreds of researchers.

This book addresses the nature of difficult problems, and techniques for addressing them. Since my own experience with such problems is centered around the development of models for engineering and management applications, a good number of the examples will exhibit a technical flavor. This is especially true for Chapters 7 and 8, which focus on research management issues. However, I would claim that much of the discussion is relevant to any domain where problems admit no obvious solutions, or even methods for identifying them.

Components of the Creative Process

The components of the creative process are depicted in Figure 1.3. The problem or task sets in motion the ideation phase. Since the solution for a difficult task cannot be obtained in a straightforward fashion, the ideation phase consists of a sequence of generate-and-test cycles: potential solutions or intermediate results are con-

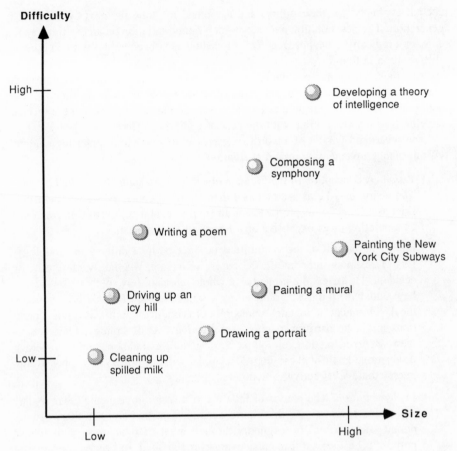

Figure 1.2. Problems characterized by the dimensions of difficulty and size.

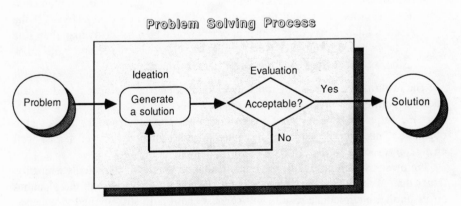

Figure 1.3. Components of the creative process.

cocted, evaluated for their utility, and examined to guide the next cycle of idea generation. The new solution may represent a minor variation on an existing candidate, or a radically new approach. This procedure usually continues until an acceptable solution is found.

At other times, the procedure may be aborted before a final resolution is reached. This may be due to the exhaustion of resources such as time, or the determination that no feasible solution exists, or the decision that the expected benefits are incommensurate with the cost of additional effort.

The successful pursuit of a worthwhile project involves four major ingredients: vision, plan, implementation, and evaluation.

- *Vision.* A conception of the ultimate objective must guide the overall project. The vision may be as abstract as a theorem or as concrete as a spacecraft; as brief as a sonnet or as protracted as an environmental policy; as self-contained as a painting or as expansive as a theory of intelligence.

- *Plan.* The pursuit of the vision must be directed by a rational methodology. Lady Luck may intervene in the course of events, whether by design or by accident. However, the project as a whole must not rely solely on chance; to be productive, it must be directed by a set of high-level plans. For example, the development of an autonomous robot to conduct search-and-rescue operations might be partitioned into the following overall strategy: literature review, experimentation, prototype construction, evaluation, and commercial production. Each of these high-level tasks will in turn be decomposed into intermediate-level activities and detailed tactics.

- *Implementation.* The plan is of little use if it is not implemented. There may be notable exceptions to this claim: a nation will discourage military or economic assault from its neighbors through a set of retaliatory plans whose purpose is best served if the plans are never activated. In general, however, a plan must be implemented to be useful. The degree of effort required to realize a vision will depend both on the nature of the objective as well as the effectiveness of the plan.

- *Evaluation*: This stage refers to an assessment of the utility of the implementation, including how well the solution fulfills its intended objectives. If the result is satisfactory, or if further attempts would be more costy than the value of the expected results, the work terminates. Otherwise the effort begins anew at the second stage with a reconsideration of the strategy.

The inspiring vision must be translated into an identifiable objective, whether it takes the form of a romantic novel, an abstract theorem, or an engineering model. Once the purpose has been identified, a strategy must be developed for attaining the objective. This overall plan may be tentative rather than definitive: it may well change in response to the efforts toward implementation.

For most difficult problems, the implementation will evolve gradually over time rather than all at once. The developmental stage often interacts with the planning effort, in that intermediate results will suggest changes in strategy and vice versa. At times, the implementation refers purely to an intellectual exercise rather than physical effort. This occurs in mathematics, for example, where a plan might involve a global strategy for proving a theorem and is followed by the implementation stage of effecting the detailed steps of inference.

When an appropriate solution has been generated, it must be evaluated to ensure its compatibility with the guiding vision. The results of the evaluation then determine whether the work should be concluded, aborted, or begun afresh at the planning stage.

Means and Ends

Much has been written in the past about procedures for solving straightforward problems, as well as creativity in the arts, sciences, and daily life. According to the traditional perspective, problem solving and creativity are distinct phenomena.

In this book, we explore a unified view of decision making and originality. In particular, we regard creativity as the hallmark of challenging tasks and therefore of intelligent behavior. An enhanced understanding of the nature of creativity and its genesis leads to rational techniques for addressing knotty problems. Against this backdrop, we will examine the strengths and limitations of conventional wisdom and traditional techniques for involving creativity. The results of this exploration can be used to develop enhanced strategies for creativity in diverse arenas, ranging from the fine arts to everyday life. For the sake of concreteness, a number of strategies for creative problem solving will be discussed in the context of research and development.

Chapters 2 to 4 will explore the nature of creativity in terms of its attributes and componential factors. An important aspect of resolving difficult problems is the relationship between resource requirements and the caliber of resulting solutions, a topic addressed in Chapter 3.

Chapters 5 through 8 focus on methodologies for addressing difficult problems. These issues range from techniques for enhancing the productivity of the problem solver to global strategies for managing the project. An important class of applications deals with challenging problems in research and development, as well as the task of supervising the first-level researcher.

Chapter 9 offers a fleeting summary of the major points and a concluding perspective.

The appendices addresses a number of interesting but more peripheral concerns. Appendix A addresses the nature of creativity in science and engineering, and presents a framework for pursuing research in a systematic way. The next addendum deals with a lighter subject—the role of creativity in humor and comedy. The third provides some background on psychological results relating to memory, a key ingredient in addressing intractable tasks. Finally, Appendix D explores a number of issues in developing a software package to enhance creative problem solving. For demonstrative purposes, some of these ideas have been implemented in the form of a prototype software package to assist in creating poetry.

But enough of promises. Shall we begin?

2

Attributes of Creativity

Innovative opportunities do not come with the tempest but with the rustling of the breeze.   Peter Ferdinand Drucker

Nature of Difficult Problems

As discussed in the previous chapter, the term *problem* is used in a general sense to refer to any task that requires resolution. These tasks may range from solving a mathematical problem to formulating a business strategy, from generating an engineering prototype to conceiving an artistic design.

A problem is called *easy* if the identification of an acceptable solution is straightforward. The label of easiness refers to the generation of the solution rather than its implementation. According to this view, finding the average value of a thousand numbers is as easy as calculating the mean of two values, since the procedure is equally straightforward.

In contrast, a *hard* or *difficult* problem is one whose resolution is not readily discernable. A common source of difficulty lies in the fact that the ultimate objective is not known *a priori*.

This situation is reminiscent of the fictional detective rummaging through a ransacked house. "What are you looking for?" asks his companion. "I don't know—but I'll know it when I find it!" In a more sedate context, the same situation applies to an investigator who wants to develop a science of manufacturing but cannot specify beforehand the nature of such a discipline. Manufacturing is one arena which until recently was regarded as a domain so complex that it would remain only an art rather than a science.

A second and perhaps more prevalent difficulty in resolving a problem relates to the route rather than the destination: the desired objective may be known, but not its means of attainment. This situation occurs when an automotive engineer must design an electric car that can travel over 1000 kilometers between battery recharges. It also occurs when a federal committee must develop a policy to contain the outbreak of a new epidemic: it is not clear to what extent emphasis should be placed on public education, medical research, governmental regulation, or other mechanisms for prevention and redress.

The resolution of such difficult problems requires a creative approach. In fact, we can summarize the preceding discussion in the following definitions.

- A *difficult* problem is a task whose resolution is nonobvious.

- A *creative* solution is a resolution to a difficult problem.

The difficulty in the project lies in the identification of the solution and/or method for its attainment. It is apparent from these propositions that difficulty and creativity are best viewed as matters of degree rather than category.

Creativity is neither the offspring of a particular domain nor the embodiment of a special process; rather it is the result of attaining special goals. In other words, it is defined solely by the product of a purposive endeavor:

> Creative thinking is not an extraordinary form of thinking. Creative thinking becomes extraordinary because of what the thinker produces, not because of the *way* in which the thinker produces it.

This view is suggested by the observation that even acknowledged masters in the arts and sciences produce works which are of uneven caliber. Hence creativity is not a characteristic of a particular person or a special process, but the outcome of a noteworthy goal and its resolution.

In other words, difficult problems engender creative solutions. Even artistic productions may be interpreted in a problem solving context. The artist's problem is one of self-expression. He has something that he wants to tell the world, or he wants to make tangible something that he thinks or feels. The purpose of an artist's efforts is to convey a personal message with depth and clarity. He wants to tell something to the world, or wants to render concrete something that moves him .

In general, creativity implies novelty but not vice versa. It is a simple task to write a computer program to produce a series of tiling patterns that might be used to serve as murals, each one differing from all its predecessors. The designs are novel but we would hesitate to call them creative. By this reasoning, novelty is a necessary but insufficient condition for creativity.

In a similar way, a child who constructs a portfolio of wildly idiosyncratic drawings will not necessarily gain much recognition from her art teacher. To be appreciated, the originality must be tempered by some coherent message, even if it is nothing more than an appeal to the observer's sense of esthetics. In other words, the novelty embodied in a product or idea must be directed toward some discernible goal if it is to be recognized as creativity.

Well-Defined and Ill-Defined Problems

Difficult problems are often ill-structured in the sense that they cannot be readily stated in precise terms. The United States is the wealthiest nation in the world by economic criteria such as the gross national product. But it is not clear that Americans enjoy the highest standard of living, as measured by parameters such as longevity or incidence of crime. If the goal is to enhance living standards in the United States, what criteria are appropriate? Here the problem itself is unstructured.

Other tasks, such as proving mathematical theorems or playing chess, are well defined. However, a problem can be stated precisely without being susceptible to ready resolution. Attaining immortality is a straightforward goal with no obvious solution.

If a problem can be resolved through a straightforward recipe, we say that the problem has an algorithmic solution. Hence an *algorithmic* procedure is one that is often guaranteed to find the best solution to the stated problem. For example, the quadratic equation

$$ax^2 + bx + c = 0$$

has an algorithmic solution known to students in high school algebra:

$$x = \frac{-b \pm \sqrt{b^2 - 4ac}}{2a}$$

On the other hand, most well-defined problems are less friendly, yielding no known algorithmic solutions. Even a simplified domain such as chess admits no surefire strategies for success. But not all is lost without algorithms. In this situation, there might be rules of thumb, or *heuristics*, that may lead to a good solution, although there is no guarantee of optimality. A heuristic in chess might be, "Control the center of the board," or in consumer marketing, "Spend 10 percent of the revenues for advertising."

Although a problem having no algorithmic solution is difficult to solve, the converse is not true. Hence a problem that admits an algorithm may still be difficult. The apparent paradox springs from the fact that the existence of algorithms is a global concept, while difficulty is local. In other words, a problem can have an algorithmic solution whether a billion people know of the recipe or only one; but the same task may be difficult to one person and easy for another. The quadratic equation is easy to solve for anyone trained in high school algebra, but is likely to be difficult for most others.

Nature of Creativity

Creativity may be regarded from a static or dynamic viewpoint. The former perspective focuses on the relationships among the components of the solution, while the latter emphasizes the sequence of events leading to the solution.

Static View of the Solution

A creative solution contains elements that exhibit both proximal and distal characteristics. The elements are close in the sense that they are linked by a common theme relating to the problem at hand. Yet by the definition of creativity, there must be at least one feature that separates them so that they are not normally perceived as neighbors in the space of concepts (see Figure 2.1).

This line of thought leads to the following proposition.

> **Multidistance Principle.** A creative solution exhibits certain features that are close in conceptual space, and others that are distant.

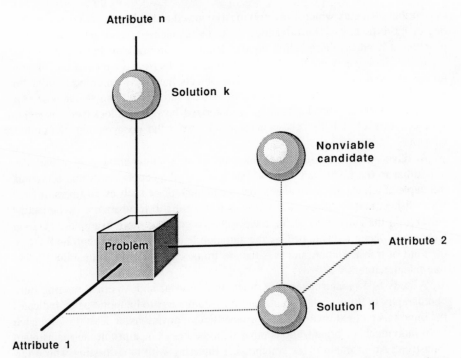

Attribute n

Solution k

Nonviable candidate

Problem

Attribute 2

Solution 1

Attribute 1

Figure 2.1. Multidistance Principle: Creative solutions are close to the problem on at least one attribute, and distant on another.

In general, the proximal aspects may be clear to the problem solver, but the distal features are more subtle. If the situation were otherwise, the solution would be obvious and therefore uncreative.

The Multidistance Principle is illustrated in the development of electronic watches. Timepieces incorporating microprocessors have largely replaced those of purely mechanical design, even though computer chips were unknown to the world only a few decades ago. A distal dimension in the two types of watches lies in their materials of composition: the heart of the traditional device consists of mechanical gears, springs and linkages, while the newer version uses silicon chips. An obvious proximal parameter is found in the constancy of repeating processes: the regularity of state transitions in both types of mechanisms is critical to the success of the product.

In the realm of humor, the static perspective is often paramount in cartoons, epigrams, and comedy. An example is found in the bespectacled child who looks up from a copy of the *Wall Street Journal* and declares, "When I grow up, I want to be a corporate raider." The humor springs from the incongruity between the speaker's age and a goal that we associate only with adults.

The Multidistance Principle has some implications for the development of software tools for creative problem solving. The tools must implement a linking strategy so that elements which were previously viewed to be distant are connected by one or more features. These features relate the elements to each other and to the problem at hand, as discussed in greater detail later in this chapter.

In developing new products, creativity often results from the recognition of new relationships among a set of objects. The objects themselves may be familiar to workers in the field, as the stock of a professional and general knowledge base. An example lies in the development of a motorized bicycle, a clock pen, or semi-synthetic textiles. Each of these items results from the juxtaposition of familiar elements.

At other times, the potential components of the solution are wholly or partially unfamiliar to the traditional experts. For instance, the copper wires that served as the staple of telephonic communications are being replaced by optical fibers, microwave dishes, and satellites—technologies that were still incubating or even inconceivable in the early days of the telephone. To cite two more examples, aviation technology was not developed by the automobile companies, although both are in the field of transportation; nor was the electronic calculator the brainchild of slide rule manufacturers.

A study of 58 major inventions drawn from Europe and America, ranging from photography to fluorescent lighting and from computers to ballpoint pens, indicates the importance of external factors. Among these innovations, at least 46 originated from individuals or organizations other than the key companies in the mainstream industries. An example is the watchmaker tinkering with brass casting who originated the process for continuous casting of steel.

Ignorance is power when it gives the novice courage to push ahead where experts fear to tread. The newcomer, bringing fresh insight and bearing no prejudice against what cannot be accomplished, transforms the impossible into the possible.

An interesting example is the development of the jet engine, an effort begun in the 1930s by two independent teams of engineers, one led by Frank Whittle in England and the other by Hans von Ohain in Germany. In both countries, most of the engineers were individuals external to the aircraft industry, or affiliated with the airframe rather than aircraft engine side of the industry. Not only did the jet engine originate from peripheral sectors, but the entire development effort received only a lukewarm reception from the builders of aircraft engines.

Almost by definition, creativity does not reside in the identification of familiar relationships among familiar objects. Hence it should not surprise us that major innovations in a field often originate from sources external to it. Outsiders are often unaware of limitations that may have hampered traditional work in a given field, but which no longer hold. Moreover, newcomers bring to bear a fresh panorama of attitudes, skills, and know-how.

Dynamic View of Creative Generation

The dynamic aspect of creativity refers to the process of generating a solution to a difficult problem. The hallmark of the dynamic phase is the fusion of alternative trains of thought in a process of positive or negative interference. The creativity consultant Edward de Bono has referred to the dynamic aspect as *lateral thinking*, to denote the fusion of alternative trains of thought. This is in contrast to *vertical thinking*, relating to straightforward inferential processes.

> Vertical thinking digs the same hole deeper; lateral thinking is concerned with digging a hole in another place. Lateral thinking seeks to get away from the patterns that are leading one in a definite direction and to move sideways by re-forming the patterns.

The confluence of separate lines of thought is a common factor in jokes. An example is found in Bertrand Russell's comment: "Most people would sooner die than think; in fact, they do so." The humor arises from a shift in perspective. In the first part of the sentence, the listener understands the word "die" in a figurative sense, but abruptly switches to a literal interpretation in the second half.

The integration of the static and dynamic perspectives is embodied in many innovative ideas. For example, Charles Darwin had been grappling for years with the empirical evidence of the evolution of species over time. This mechanism appeared comparable to the deliberate breeding for characteristics such as high yield in grain or stamina in racehorses. Such characteristics were developed under human guidance. Unfortunately, Darwin was unable to account for the agent of change in natural evolution. One day he was reading a book on political economy describing the culling effects of war and pestilence on human populations. By merging these two lines of thought, Darwin postulated the principle of the survival of the fittest as the guiding factor in natural evolution.

In this example, the merging of the concepts of population selection and natural evolution embodies the static view of creativity, which focuses on the fusion of disparate concepts. This result arose from the confluence of two lines of thought— one in biology, the other in social theory—a process that illustrates the dynamic perspective.

Components of a Solution

The solution to a difficult problem rarely arrives full-blown, but rather in pieces. We will refer to these elementary concepts or items of intermediate insight as ideas.

- An *idea* is a concept that may serve as a component of a solution.

The idea may relate to some aspect of the solution or to an intermediate insight that ultimately helps to yield the solution.

Consider the problem of bringing home souvenirs from a vacation when the outgoing trip had already begun with a set of overpacked bags. The idea of mailing home part of the returning baggage—say, a package of books or clothes— represents a solution to the problem.

On the other hand, consider the problem of reviving a company battered by heavy competition. The strategy of closing down production plants for a month to cut labor expenses and slash inventories is an idea that might constitute part of the solution. Another idea may be to sell off a peripheral division to generate cash flow, or to increase sales by cutting prices. A strategy of long-term consequence is to introduce new products by increasing research expenditures. All these ideas serve as components of the solution to the problem of reviving an ailing company.

An idea may itself depend on other ideas. For example, the strategy of generating new products may be based on the concept of conducting telephone interviews with established customers, or on a broad assessment of technologies under development in university research centers. In turn, these ideas may depend on the notion of hiring consultants to assist in the design of a questionnaire or the formulation of a coherent technological strategy.

Ideas may be classified as elementary or compound notions depending on their atomicity. An *elementary* idea is one that cannot be reasonably partitioned into smaller concepts. An example lies in the use of a wedge to force open a locked door. A wedge represents an elementary idea, as does the concept of opening a door and the notion of leverage. These elementary ideas together constitute a composite concept that serves as the solution to a problem. In this way, a *compound* idea is one that consists of other ideas, each of which may in turn be elementary or compound.

An elementary idea is a simple object that either exists or does not; it springs from the mind of a particular agent, whether a person or a component of a computer program. A group of individuals may inspire ideas among themselves, but each atomic idea is given life by a specific agent. This stipulation does not preclude the generation of the same idea by different individuals, as often happens in science and art, as well as in everyday life.

The distinction between elementary and compound ideas will be explored in greater depth in later sections, including the prescriptive implications of the distinction.

Evolution or Revolution?

Do great ideas arrive fully grown, or do they evolve gradually over time? The answer seems to be both, depending on the observer's view of what constitutes a seminal idea.

The argument has been made that major advances in science occur through a series of novel paradigms or models. Examples of such paradigm shifts are the transition from a geocentric to a heliocentric view of the heavens, and the conservation of energy in its multifarious forms. The paradigms are viewed as sudden or revolutionary phenomena, each of which is followed by a quiescent period during which scientific advances occur incrementally, by filling in the gaps in the paradigm and perhaps extending it in minor ways.

But what is the nature of the development process *during* a revolution that marks the birth of a paradigm? Does the paradigm or idea arrive in full regalia at dawn one fine Sunday morning? This does not seem likely. In the words of the behavioral scientist Herbert Alexander Simon:

> Most scientific activity goes on within the framework of established paradigms. Even in revolutionary science, which creates those paradigms, the problem and representations are rooted in the past; they are not created out of whole cloth.

It is convenient for our purposes to distinguish between elementary ideas and compound ones. The solution to many a difficult problem lies not in a solitary idea, but in a collection of them. For example, the science of thermodynamics rests on two basic postulates: the conservation of energy and the tendency toward disorder. These elementary ideas, central as they are, do not form the whole of thermodynamics; they are but two pieces in a storehouse of ideas.

Let us return to the story of Darwin's seminal insight. The book on social dynamics that led him to the theory of natural selection, *Essay on Population*, was written by the English economist Thomas Robert Malthus. The book argues that the growth rate of any population tends to exceed its supply of food; hence the fittest members triumph in the competition for food, while their unfortunate siblings wither away. The surviving members then bequeath their superior characteristics to their offspring, thereby providing a mechanism for the eventual evolution of a variant species.

Although Malthus' essay was a catalyst for the emerging theory of natural selection, it was only one item in a longer chain of events. Darwin had come to read the essay in October 1838, fifteen months after he had initiated a systematic investigation of species formation. Moreover, Darwin's initial reading of Malthus' book did not have an immediate impact on him. It was not until four years later, while traveling in a carriage, that the idea of natural selection suddenly occurred to him.

Even more important, the time was ripe for such a theory to evolve. In Europe, the scientific community was beginning to wonder about evolutionary phenomena and provocative evidence such as fossils of extinct animals. In addition, artificial breeding—the selective breeding of plants and animals for specific traits—was known to Europeans even before Darwin's age.

The evolutionary nature of ideas has been characterized by Douglas Hofstadter, a computer scientist, as variations on an underlying theme:

> The crux of creativity resides in the ability to manufacture variations on a theme.
> . . . If you look at the history of science, for instance, you will see that every idea
> is built upon a thousand related ideas. Careful analysis leads one to see that what
> we choose to call a new theme is itself always some sort of variation, on a deep
> level, of previous themes.

The cognitive psychologist Jay Paul Guilford voiced a kindred view when he asserted that no idea is 100 percent novel.

An elementary idea may occur suddenly, but a compound idea that consists of a collection of smaller ideas is likely to evolve over time. This line of thought is pursued further in Chapter 5 on the output phase of idea generation.

Studies in creative problem solving indicate that innovative solutions usually evolve incrementally through a series of small ideas rather than in a single stroke of inspired genius. This is true both in laboratory settings as well as in postmortem analyses of masterworks in diverse fields: Francis Scott Fitzgerald in literature, Pablo Picasso in art, and Johann Sebastian Bach in music, to name a few.

Such artistic creations are the culmination of many hours or even years of effort, forming gradually in incremental fashion rather than as a single complete piece. The evolutionary nature of such effort is apparent in the fact that the creators usually expend much time and effort in false starts, sometimes abandoning the initial line of development or adopting completely radical approaches to express their themes. Even when the artist settles upon the final approach, he or she usually incorporates numerous revisions and minor modifications.

We see that elementary ideas are by definition discontinuous phenomena. However, difficult problems usually demand a constellation of elementary ideas and thereby evolve over time. The nature of the solution, as well as its period of incubation, will depend on the character and magnitude of the engendering problem.

Summary

Creativity is defined by a problem or task, rather than a discipline, person, or process. As a result, creativity is a phenomenon of degree rather than category. It may be found in the conception of a sculpture, the composition of a symphony, or the removal of a stubborn stain.

- A *difficult* problem is a task whose resolution is not obvious.
- A *creative* solution is a resolution to a difficult problem.
- Creativity may be regarded from static or dynamic perspectives.
- The static view is encapsulated in the Multidistance Principle: a creative solution contains elements which are close in conceptual space as well as those which are distant.
- The dynamic view refers to the process of generating a solution.
- An idea may by elementary or compound.
- A solution consists of one or more elementary ideas and tends to evolve over time.

3

Quality of Ideas
and Solutions

The difficult we do immediately. The impossible takes a little longer.
Slogan of the United States Armed Forces

Quality of an Idea

We pass judgment on the value of scientific results. For example, we may feel that the periodic table of elements is of greater significance than the understanding that hydrogen is lighter than oxygen. For one thing, the latter fact may be deduced from an understanding of the periodic table. In this way, we may ascribe a measure of quality to an idea.

The quality of an idea is subjective, and cannot be assigned an absolute value. However, it is possible to give a partial ordering and claim that idea *A* is of higher, lower, or equal worth in relation to idea *B*. The judgment of the relative importance of ideas is made routinely, for example, by an instructor in delivering a lecture or writing a book. The assignment of value occurs implicitly in the selection of topics and their relative emphasis.

Once we admit a preference ordering among ideas, we may also assign an arbitrary numerical scale to them. This practice is standard in the field of economics, where a preference ordering among goods suggests a measure of utility. Since each consumer has individual tastes and needs, the resulting utility function varies from one person to another even for the same basket of goods.

Further, the preferences are subjective and relative, rather than absolute. As a result, the level of utility can be based only on a conceptual scale. The basic measure of utility is an arbitrary unit called a *util*.

In a similar way, we may assign a quality metric in terms of a granular unit of a *qual*. A person may assign a particular set of quals to a portfolio of ideas based on his own tastes and predilictions. A second person may offer a completely different set of quality values. This conception of an individual ordering of ideas is consistent with the view of difficulty and creativity as relative rather than absolute parameters.

To pursue this line of reasoning, we may also speak of the combined quality of two or more ideas. The value of a set of ideas may be greater than, equal to, or less than the sum of the individual values.

- *Greater*. Two or more ideas may imply further consequences. Suppose that, the rule, "If the barometer falls, the weather will turn foul" and the fact "The barometer is falling" are known. These items, in conjunction with the rules of logical deduction, imply the further fact "The weather will turn foul."

- *Equal*. The set of ideas is independent: no idea interacts with any of the others in a positive or negative way.

- *Less*. The set of ideas is redundant: at least one idea is a logical consequence of the others. An example was given earlier in the context of the periodic table.

Another example relates to the transitivity property of algebra. One form of the transitivity principle relates to inequality: $A > B$ and $B > C$ implies $A > C$. In this domain, the set of items $\{Transitivity, A > B, B > C \}$ gives rise to the additional fact $A > C$. On the other hand, $\{Transitivity, A > B, C > D \}$ will yield no further results, since the second and third facts contain no common elements on which the transitive property can operate. Finally, the set $\{Transitivity, A > B, A > B \}$ is redundant, and therefore provides no more value than a comparable one that lacks the second or third item.

The total value of a portfolio of ideas, then, is given by some function of the individual qualities of the component ideas. In this way we may speak, for example, of the quality of a line of research having greater or lesser value than that of another.

Effort and Quality

The expected quality of a solution seems to depend in a continuous way on the amount of input effort. The expected value of an elementary idea, however, is a little more involved. The main difficulty is that an idea is a discrete beast that either exists or does not. A solution may be generated by a committee, but an idea springs only from the mind of an individual.

We do not yet understand how the human brain creates sonnets, composes symphonies, and performs leaps of intuition. We engage in intuitive acrobatics when we recognize patterns with only an incomplete set of facts. Although many forms of mentation remain a mystery to us, one thing seems clear. Mental feats, great or small, take time for their realization.

This temporal requirement may be explained at the microlevel in terms of electrical and chemical processes involved in neural activity. These processes transpire over time, and the delays eventually translate into higher-level requirements at the cognitive level. Psychological experiments indicate that sensory stimuli such as visual input must persist for at least 10 milliseconds before they are properly registered in the human sensory system. The likelihood that a stimulus will be perceived correctly—or that it registers at all—increases with the duration of the input signal.

When any form of physical response is required, the minimal threshold increases by an order of magnitude. For example, a driver requires at least 100 milliseconds before the recognition of dangerous road conditions is translated into pressing the brakes.

According to one estimate, the brain seems to require roughly 10 seconds to encode one "chunk" of information into memory. An example of a chunk is a friend's birthday or the fact that oxygen is involved in rusting.

The trend is clear: the time required to perform intelligent actions tends to increase with the complexity of the task. This generalization seems to hold even within the subdomain of ideas. Ideas of greater value tend to require more time to generate—and later to refine—than simpler ones.

Time is a necessary, but not sufficient, factor in developing good ideas. All of us know people who propose new ideas with surprising infrequency.

Although time is a critical dimension in solving problems, other parameters exist. Among these are resources such as money, energy, or additional minds. These resources may influence each other positively, negatively, or not at all, depending on the problem and its solver.

The relationship between the quality of an idea and the effort for its germination is depicted in Figure 3.1. The northwest frontier, which defines the upper periphery of points on the chart, embodies the concept of conditional output based on the level of input.

Minimal Effort Principle. Effort is a necessary but insufficient determinant of the quality of an idea.

This concept is illustrated in the observation that good ideas take time to develop. For easy problems, there may be a strong correlation between the amount of time spent and the achievements won. The figure suggests, however, that a problem solver may work indefinitely on a difficult project without obtaining an acceptable solution. We may state this as a corollary of the Minimal Effort Principle.

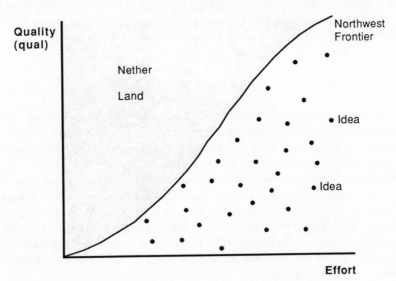

Figure 3.1. The quality of ideas versus effort (time, energy, money, etc.).

**Quality
of Idea**

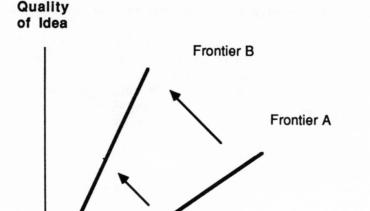

Figure 3.2. Northwest frontiers of differing slope.

Corollary. The amount of time spent on a task is not a sufficient measure of progress.

The slope of the northwest frontier is arbitrary, since the vertical axis is given in units of quals. It is clear, however, that the slope is higher for certain individuals than for others, as depicted by the two frontiers in Figure 3.2.

It is also clear that a person can change the slope of her own northwest frontier through conscious action. This concept is discussed further in Chapter 5 on the subject of techniques for enhancing productivity in problem solving.

Input Power vs. Output Quality

The amount of effort expended over time might be called *power,* in keeping with the terminology used in the physical sciences. Since an idea is a discrete object that either arises or does not, there is little sense in speaking of the relationship of a *particular* idea to the rate of input effort.

On the other hand, we may speak of the probability of obtaining an idea of given caliber as a function of the input effort. This three-dimensional structure is depicted in Figure 3.3. The figure indicates that there is some lower threshold P_{min} below which no useful ideas will occur, and conversely an upper bound P_{max} above which the problem solver will be paralyzed by fatigue. The ridge across the landscape in the figure indicates that, between the lower and upper thresholds, the likelihood of obtaining a high-caliber idea rises with increasing power.

We should keep in mind that difficulty and creativity are local rather than global

Probability   **Quality**

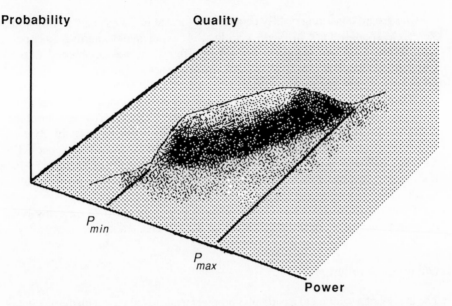

P_{min}

P_{max}

Power

Figure 3.3. Likelihood of obtaining an idea of specific quality as a function of the input power (rate of effort).

parameters. Since they depend on the individual problem solver rather than some absolute scale, the exact location and shape of the ridge in Figure 3.3 will vary for each person and problem.

In addition, the precise shape of the trade-off between input power and output quality depends on the nature of the effort—be it time, energy or another parameter. If the effort is measured by time, then we have a dimensionless quantity P having units such as hours per hour. In this case, the lower threshold might be only a few seconds for performing arithmetic calculations, or several hours per week for an obstinate research problem. The quality of output will again vanish for high values of P.

The existence of a lower threshold P_{min} may appear obvious. But sometimes we ignore the obvious. For example, this oversight may be the single major roadblock for graduate students involved in research: some take three or four years to complete a master's program while being registered as full-time students, and others take a decade to finish a doctoral dissertation. Still others never graduate at all. These people usually fail to recognize the nonlinear relationship between effort and results. Many of these individuals would complete their research much sooner—and sometimes at all—if they were to increase their rate of effort above the lower threshold.

The existence of a useful upper limit on input power applies not only to individuals but to entire groups. In 1976, the Data General Corporation established two engineering teams to develop the company's new-generation computer. Unlike previous machines based on information packets 16 bits long, the new minicomputer would depend on a 32-bit architecture.

One group, consisting of fifty people, was located in a plush new facility at the Research Triangle Park in North Carolina. After an initial abortive attempt, a smaller group of five engineers headed by Tom West was established next to corporate headquarters in Westborough, Massachusetts. The smaller project had lukewarm support from management, who considered it largely a waste of effort.

When West's team completed the new machine in the spring of 1980, the group at North Carolina still had little to show for their efforts. The new machine, christened the Eclipse MV/8000, was announced to the world on April 29. By early 1981 the machine accounted for perhaps 10 percent of the company's revenues. The computer had arrived just in time to bolster the upper end of Data General's product line, an area that had grown anemic due to heavy competition from other manufacturers.

This story illustrates the point that might does not always yield right in solving difficult problems. A power drill cannot be used to create a tapestry.

Quality of a Solution

The quality of a solution to a particular problem tends to increase with the amount of effort expended, as shown in Figure 3.4. The effort may be in terms of time, money, energy, or some other parameter.

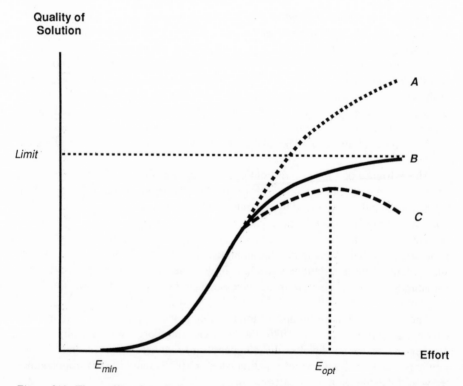

Figure 3.4. The quality of a solution as a function of effort.

The figure indicates that there is some minimal effort E_{min} below which a solution, if one materializes at all, will be of little or no use. This lower threshold will obviously depend on the application. It may be a fraction of a second for a driver to avoid hitting a child darting across the road, or years for resolving an unsolved problem in theoretical physics. In the financial realm, the lower threshold may be a few pennies to glue a broken ruler, or billions of dollars to land a human on Mars.

The quality of a solution often increases with the amount of effort expended, as shown in curve *A* of Figure 3.4. At other times, the quality of the solution will level off and converge toward some asymptotic value, as depicted by curve *B*. For example, it is unlikely that the Manhattan Project would have developed a significally better atomic bomb within the same time frame, even if it had been supported by a budget two or even ten times as large. The stork requires nine months to deliver a baby, no matter how anxious the parents.

Time constraints have a negative impact on productivity in creative work. This observation is consistent with the view of decreasing marginal returns: the quality of a solution rises with an increase in the total time dedicated to resolving a problem, but the rate of this increase actually decreases.

There are other times when the quality of the solution reaches a peak at some optimal value of effort E_{opt}, then begins to fall. This was a common experience in large-scale software development efforts in the 1960s, when small teams often developed better programs in less time than larger ones.

Throughout this period, difficult software projects were behemoths without precedent. Having no prior experience to guide their efforts, programming teams groped for any tactic that might improve their plight. One obvious tactic was to throw more resources at the problem; but power *in* did not equate to progress *out*.

> More software projects have gone awry for lack of calendar time than for all other causes combined . . . when schedule slippage is recognized, the natural (and traditional) response is to add manpower. Like dowsing a fire with gasoline, this makes matters worse, much worse.

Today, software development is a much more tractable task, at least in the arena of conventional information systems. In the 1970s, techniques such as structured programming became widely available to guide the development process.

But the story differs for advanced software, such as expert systems which incorporate artificial intelligence techniques. The technology of knowledge-based systems did not begin to mature until the early 1980s. As a result, our collective experience in constructing knowledge systems lags behind that of conventional software by a couple of decades. Since few precedents are available, the development of an advanced system is still an uncertain affair highly susceptible to failure.

Quality vs. Quantity

One cannot determine *a priori* the quality of a solution designed to address a particular problem. This view is supported by the observation that creative individuals produce works of varying caliber, and the most renowned figures in any field of endeavor are usually its most prolific contributors. The probabilistic nature of the quality of creative output holds longitudinally over the lifetimes of specific individuals, as well as cross-sectionally across diverse realms of science, art, and politics.

> Certain prolific persons are responsible for a disproportionate share of the achievements in any given endeavor and . . . this quantity of productive output is probabilistically connected to quality of impact, or eminence. . . . Though there seems to be a definite productive peak, the constant-probability-of-success model operates for both longitudinal and cross-sectional data, signifying that the odds of a creator's conceiving a quality product are always proportional to the quantity of products, the creator's age notwithstanding.

This observation relates to the collective quality of solutions that address separate problems. It does not imply, however, that quantity yields quality for a particular task. It is easy to generate any number of potential solutions of low quality in response to a problem, and to end up with mediocrity despite quantity.

This is precisely the limitation faced by haphazard methodologies such as brainstorming. In its orthodox form, brainstorming calls for the rapid generation of ideas without concern for their relevance or quality. Although numerous ideas are generated, the majority of these are of dubious value. As a result, for many difficult arenas such as technical research, wanton brainstorming is seldom used. We will turn to these topics in greater depth in Chapter 5 on methodologies for individual problem solving.

Summary

The quality of an idea, being a subjective characteristic, cannot be assigned a definitive value. However, an arbitrary scale whose units of measurement are defined to be quals may be assigned. The value of a set of ideas can be defined in the same way, its collective utility being greater than, equal to, or less than the sum of the individual values.

- By the Minimal Effort Principle, effort is a necessary but not a sufficient determinant of the quality of an idea.

- A corollary to this principle is that the amount of time spent on a task is not a determinant of progress.

- The northwest frontier defines the maximum quality of ideas as a function of input effort.

- The northwest frontier can be shifted upward through conscious effort and the use of effective strategies.

- The quality of a solution to a particular problem tends to increase in a probabilistic sense with the amount of input power, then may decrease after exceeding a threshold rate.

- The quality of an idea cannot be determined before it is conceived. Hence the likelihood of obtaining a good solution increases with the number of promising candidates that are generated.

4

Factors of Creativity

When inspiration came, it was in the form of a picture, a mental image of two small wavy forms and one big one. That was all—a bright, sharp image etched in his mind, no more, perhaps, than the visible top of a vast iceberg of mental processing that had taken place below the waterline of consciousness. It had to do with scaling, and it gave Feigenbaum the path he needed.

James Gleick

The dictionary defines a *factor* as "something that actively contributes to the production of a result." A number of factors may be attributed to the creative process and its final product. These factors are purpose, diversity, relationships, imagery, and externalization. The ingredients of creativity seem to be the same for diverse domains, from the arts to the sciences and social studies.

The factors of creativity are depicted in Figure 4.1. These parameters refer to the purpose of the project, plus four others identified by diversity, relationships, imagery, and externalization. The second and third factors define the structure of the problem and its solution, while the last two relate to representation issues for generating the solution and expressing it in some form.

The creative factors are listed in Table 4.1, along with their defining characteristics and prescriptive implications. The operational implications are described for both the human problem solver, as well as for the computer system that might be developed to assist in this task. For example, imagery relates to the development of ideas through sensory mechanisms, whether in actuality or in conception. The prescriptive implication is to generate a series of images in various formats, whether pictorial, auditory, or tactile. Of these, the most powerful vehicle is the visual image which can simultaneously represent numerous objects and their relationships. The operational implication for a computer-based system is a rich store of icons or pictorial images that may be depicted on color screens using versatile graphic techniques.

Purpose Factor

The problem to be resolved defines the purpose of the creative process. The factor of purpose involves the distillation of a problem into its essential elements. This

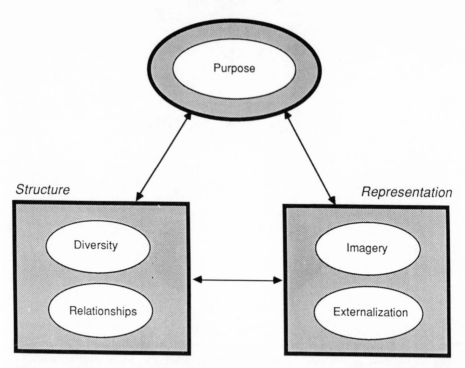

Figure 4.1. Factors of creativity. Diversity and relationships refer to the structure of the solution, while imagery and externalization denote representational factors. All four factors support the guiding purpose.

involves the identification of critical attributes and the elimination of extraneous features.

For millennia people had yearned to fly. The existence of birds was proof that flight was possible. Man's straightforward approach to achieving flight was to imitate aviary shapes and motions, especially wings and their flapping. But birds have functional requirements other than airborne motion, such as feeding, fleeing, and reproduction. It was not until the peripheral activities such as wing-flapping were ignored, that man could design a machine allowing him to attain active flight. Today the airplane flies higher, faster, and farther than any of nature's creations.

Identifying the purpose of a problem solving effort enhances productivity in generating effective solutions. Psychological studies confirm that the quality of solutions tends to increase with a better understanding of the criteria for their evaluation. In other words, a problem solver generates more effective solutions when armed with prior knowledge of the ways in which those solutions will be judged.

Problem solving in the real world is complicated by its dynamic character. The objectives often form a moving target shaped not only by the changing perceptions of the problem solver during the process of resolution, but also by the environment, whether independently or in response to an intermediate problem-solving step.

Table 4.1. Factors of creativity: descriptive characteristics plus prescriptive implications for enhancing each factor

Factor	Description	Prescription	
		Human	Computer
Purpose	Formulation of objectives. E.g., purpose of a flying machine is to transport by air, not to flap wings.	Determination of key goals.	Repertoire of key objectives, for limited domains. Interactive procedure to help user isolate critical features.
Diversity	Fusion of disparate objects. E.g., Boolean logic to transform arithmetic operations into reasoning procedure in computers.	Pursuit of diverse, seemingly unrelated paths of inquiry. Enhancement of memory.	Large secondary storage to complement human memory. Access to commercial databases and archives.
Relationships	Patterns among objects, and their effective representation. E.g., Analogy; insight through parallel constructions.	Recognition of similarities and differences. Knowledge of history.	Alternative representations. E.g., Graphs vs. charts. Repertoire of analogic relationships. Knowledge base of related problems. Rapid enumeration of alternatives.
Imagery	Ideation through images, whether visual, auditory, or tactile. E.g., curling snake to suggest benzene rings.	Visualization: Generation of tangible and intangible images. Recognition of complex relationships through pictures. Relaxation after concentration: ideation through dreams and reveries.	Repertoire of icons for objects and their potential relationships. Graphic techniques.
Externalization	Ideation and communication through tangible and intangible channels. E.g., creation of sculpture; construction of a model; composition of poetry; formulation of a cogent argument.	Auto-feedback through externalization; Writing notes, conducting monologues, or explaining/teaching. External feedback from other individuals.	Communication through voice recognition and speech generation. Reading documents and preparing written reports.

The case of changing perceptions occurs when a sedentary worker recognizes that his efforts to improve job performance are not his real problem. Rather, a better task definition is to find a mobile job catering to his need for physical activity. The use of sailing ships as military vessels illustrates the situation of an exogenous development affecting a problem definition. In particular, a program to increase the speed of sailing ships is rendered obsolete by the advent of steam engines and aircraft technology. The problem of external factors becomes especially acute in the face of intelligent adversaries. A marketing organization must reconsider its strategy of reducing prices to promote sales if its competitor cuts prices even lower for corresponding products.

Motivation

The purpose of a problem-solving activity can usually be viewed at many levels. The basic level is the objective of resolving the task itself, whether it be to pen a sonnet, prove a theorem, or earn a billion dollars.

However, solving the problem may only be one component of a higher-level objective. A sonnet may constitute a stepping stone for becoming an acclaimed poet; a theorem, one part of a doctoral dissertation; and a billion dollars, a means to philanthropy. Whatever the ultimate objective, performance of the task at hand may be affected for better or for worse by motivational factors.

The American psycholgist Catharine Morris Cox conducted a study of 301 historical figures and reached the following conclusions regarding the importance of personality:

> All equally intelligent children do not as adults achieve equal eminence. . . . Youths who achieve eminence are characterized not only by high intellectual traits but also by persistence of motive and effort, confidence in their abilities, and great strength or force of character.

The importance of motivation is as relevant to success on a specific problem as to success in life in general.

Diversity Factor

Creativity involves the association of objects that previously had appeared disparate. In this sense it resembles humor, which has been described as the "juxtaposition of incongruous elements." The relationship between creativity and humor is discussed at greater depth in Appendix B.

Organizations tend toward routinization in order to increase efficiency in processing information for recurrent events. Unfortunately, the prevalence of standard operating procedures tends to shackle innovative behavior. A major tool to promote creativity is to develop relatively autonomous units such as those in research and development, strategic planning, and new-product marketing.

The role of diversity in creative problem solving was discussed extensively in Chapter 2, in connection with the nature of the solution. According to the Multidistance Principle, a feasible solution must exhibit both distal and proximal characteristics in relation to the originating problem. The first part of this principle embodies the diversity factor by identfying conceptual distance as a requirement for candidate solutions.

The second half of the principle concerns the relationship among the disparate components of the solution, some aspects of which must lie near each other in conceptual space. This proximity is defined in terms of the relationship that links the component ideas. We now turn to the relational dimension.

Relationship Factor

Creativity involves not only the identification of diverse elements, but their fusion into a synergistic whole through a set of relationships. Past experience may help us identify such relationships through positive or negative examples. A large knowledge base is a necessary but insufficient condition for a productive mind. In fact, a knowledge base cluttered with too much trivia may impede deliberate reasoning, let alone creative thought. This predicament is dramatized in the movie *Paper Chase*, in which a law student possesses a photographic memory but is unable to identify the salient points of a case.

Some common relationships are listed in Table 4.2. The first entry, for example, indicates that the *equality* relationship refers to two objects that are identical or equivalent; one-tenth is the same amount as 10 percent, while the sun is the same object as our closest stellar neighbor.

Representing a problem in alternative modes of expression is often key to its resolution. The alternative modes may take both physical and abstract form. The techniques range from the verbal diagnosis of the physician to the scale model of the architect, and from the differential equations of the physicist to the computer model of the economist.

Often the proper representation of a problem leads to its immediate solution. This is epitomized by algebraic word problems such as, "If Al is twice as old as his mother when she was four years younger than Bo, and Bo at the time was six years younger than Al is now, how old is Al?" Most of us would find this problem difficult or impossible to resolve mentally. But with a few tricks from algebra we

Table 4.2. Examples of relationships

Type	Description	Examples
Equality	Objects are identical or equivalent.	0.1 : 10 percent. Sun : Closest star.
Similarity	Objects resemble each other in form or function.	Two congruent triangles. Pen : Pencil.
Contrast	Objects are opposites.	True : False. Isolation : Communication.
Membership	One is an element or subset of the other.	1 : Integers. Mammals : Animals.
Composition	One is a part or component of the other.	Letter : Word. Branch : Tree.
Utility	One employs the other.	Arithmetic : Algebra. Subroutine : Program.
Causation	One induces the other.	Moon : Tides. Worry : Ulcer.

can readily determine that Al is twenty and his mother was ten when Bo was fourteen.

Mathematical and verbal representations together form a more powerful tool for solving problems than either one alone. More generally, the appropriate choice of representation tools can be the most important step to solving a challenging problem.

Linkages

Associations refer to established connections among ideas. The creative process requires the forging of new links among disparate concepts, a process which has been called *bisociation*. The invention of the printing press is a good example of this. Gutenberg was familiar with the notion of forming symbols on small balls of wax, a practice used to impress emblems in sealing and authenticating correspondence. In a similar way, an entire page of symbols might be formed all at once by using a sizable plate to press a larger ball of wax. Unfortunately, the large plate would be difficult to control, since one corner or other would invariably sink lower than the others and thereby distort the pressed image. The solution to this problem came to Gutenberg one day while he was attending a wine fair: he observed a wine press, an apparatus that uses a flat surface to apply a uniform force over a wide area. In this process, the bisociation of seals and wine presses led to the conception of the printing press.

Analogy

An analogy is a similarity between two relationships. Since it deals with the relationship between relationships, it is a second-order concept. The relationships in an analogy are equivalent by virtue of being similar or even identical. For example, the moon is similar to an apple, since both are attracted to the earth by gravity. Or a horse is analogous to a car in that both can serve as transportation mechanisms. Consider the observation, "An apple is to fruit as a salmon is to fish." This statement asserts that the first pair of objects resembles the second pair in that the first item in each is an example of the second. In a similar vein, "Energy is to physics as utility is to economics" alludes to the fact that the first item in each pair is a central, unifying theme in the study of the latter.

The past provides analogies that may serve as points of departure for new ideas. Louis Pasteur is said to have noticed that grapes ferment only when their skin has been broken. From this he inferred that infections are caused by external agents rather than internal factors.

Another analogy inspired Thomas Edison in his design for the kinetoscope, a forerunner of the movie projector. He had already invented the phonograph to record audio signals, and this new project was a deliberate attempt to develop an analogous device for recording images.

One methodology within the analogical approach is to solve a simpler problem when the original task appears too formidable. This happens, for example, when a student of algebra is confronted with the problem of obtaining solutions to the equation

$$x^4 - 5x^2 + 4 = 0$$

The solution to the problem is not readily available. However, transforming the equation with the substitution $y = x^2$ leads to the simpler problem of solving

$$y^2 - 5y + 4 = 0$$

This is an elementary problem in high school algebra, to which the solutions are $y = 1$ and $y = 4$. At this point, the substitution of x^2 for y leads to two equations of even greater simplicity: $x^2 = 1$ and $x^2 = 4$. The two pairs of solutions yield the values of $+1$, -1, $+2$, and -2 which satisfy the original problem. In this way, a difficult problem can be resolved by addressing a series of easier problems.

Past as Prologue

Routine problems, by definition, are those that have a sufficient number of precedents to suggest a ready solution. The accumulated knowledge may be informal, residing in the minds of experienced individuals. Or it may be formalized, having been stated explicitly in the form of standard operating procedures, scientific laws, or software packages.

Difficult problems, on the other hand, are rare or even unique, and as such have few precedents to guide their resolution. For this reason, history and personal experience are of less relevance than they are for routine problems. Yet even here, the past is a basis for the future.

For 13 days in October 1962, the world stood on the brink of nuclear holocaust. The United States had obtained photographic evidence of the construction of medium-range nuclear missile sites in Cuba, apparently under Soviet direction. President John Kennedy demanded the dismantling of the facilities and the removal of all Russian nuclear missiles from Cuba. But the Soviet leader, Nikita Sergeyevich Khrushchev, denied that any offensive weapons were being deployed.

This was a crisis of unprecedented significance, for never before had two superpowers faced each other under the threat of nuclear war. Both sides were well aware that if the crisis were to escalate, it could end in a conflict that could destroy the two nations, and perhaps even obliterate western civilization.

The President and his aides considered a number of options. Among these was direct military intervention in Cuba: a surprise attack against the missile sites and perhaps other military bases as well. However, this alternative brought to mind the Japanese attack on the United States in 1941; the prospect of a "Pearl Harbor in reverse" seemed untenable.

A second option was to do nothing. The Kennedy administration was weakened by the Bay of Pigs incident in 1961, when a small-scale invasion of Cuba had turned into a fiasco. If Kennedy were to appear indecisive during the missile crisis, his administration might well collapse due to disenchantment at home for incompetence and criticism from abroad against lack of American resolve. Moreover, the appeasement of Soviet belligerency in the 1960s was, in the long run, not likely to work any better than it did for German expansionism in the 1930s.

The third major option, a naval blockade, would represent a controlled demonstration of resolve. The U.S. Navy would establish a line of quarantine around Cuba and board all Russian ships crossing it.

This last option was selected. Even as the blockade was implemented, Kennedy was aware of the lessons of World War I: a series of mistakes and accidents could overtake the course of events and lead the two nations into full-scale conflict, despite restrained measures by their respective leaderships.

Throughout this incident, the past served to guide present action for future purpose. This was done by delving into history to cull comparable events and exploring not only their similarities in the past, but also the differences.

In the process of scientific discovery, for example, the past can guide the present but only loosely. Isaac Newton was said to have been fortunate as well as brilliant: there is only one system of the world to be discovered, and he was there at the right time.

The history of science suggests a series of stages for the development of a discipline. First is the construction of a *framework*, consisting of a core set of concepts. Some examples include matter, energy, waves, and particles in physics; utility, services, and markets in economics; and memory, processes, and computation in computer science.

The second stage involves the development of *models* to depict basic concepts and their relationships. These are often couched in the language of mathematics.

The third stage relates to the behavior of the basic objects under various conditions. Examples are mass-energy conservation in physics; the elimination of long-run profits in equilibrium economics; and the time-space trade-off for computer algorithms.

These parallels would suggest a similar approach to the development of a science of manufacturing. The first step is to consolidate existing knowledge into a set of key concepts, then to develop a set of mathematical tools to state these concepts precisely, and finally to formalize the relevant operating principles. As with most complex tasks, these steps are only partially independent. For example, the drawbacks of representing certain operating principles in, say, linear algebra, will suggest the use of other mathematical models.

The sequence or order of objects relative to each other is often key to the creative solution. The celebrated mathematician Henri Poincaré has written of the importance of unifying relationships in his field:

> A mathematical demonstration is not a simple juxtaposition of syllogisms, it is syllogisms *placed in a certain order,* and the order in which these elements are placed is much more important than the elements themselves. If I have the feeling, the intuition, so to speak, of this order, so as to perceive at a glance the reasoning as a whole, I need no longer fear lest I forget one of the elements, for each of them will take its allotted place in the array, and that without any effort of memory on my part.

In mathematics, as in other fields, the high-level objective determines the requisite structure at all levels.

Imagery Factor

The language of words is a powerful means of ideation and our most versatile tool for communication. This is true for the written as well as the spoken word.

However, verbal language is a relative newcomer in evolutionary terms. In all likelihood, this tool—as a versatile mode of expression—is not much older than our vertebrate ancestors who appeared on the globe some hundred thousand years ago.

Before words appeared, when our humble ancestors were still roaming the oceans, images were a key source of knowledge acquisition and dissemination. Often the most useful images were those obtained at a distance, away from harm's way: vision, audition, and olfaction. The state of the environment was perceived through images, and any deliberation about fight or flight was probably effected through similar media. To the extent that information was processed, it was likely to express itself in the language of images.

Given the obvious importance of rapid object recognition and environmental awareness in the quest for survival, it is not surprising that image processing is a highly developed facility among animals, including man. Even today, a picture can paint a myriad of words, and verbal communication is itself often conveyed through the printed image.

Many of our ideas are transformed into images and probably encoded as such. For example, when we recall a novel read the previous year, we recollect the plot as a sequence of images rather than a string of words used by the author.

Images are central to creative ideation, as many of our inspirations occur in pictorial form. Some novelists, for example, assert that their plots unfold in the mind's eye, and then are simply recorded on paper.

According to legend, René Descartes invented the rectangular coordinate system after watching a fly walk across his ceiling. He wondered how the position of the fly could be determined in a simple way, and concluded that one pair of numbers, each representing distance along an independent axis, would suffice for the task. In a similar example, Albert Einstein's ideas on relativity were supposedly due, in part, to his visualization of the "experience" of riding on a photon in its traversal through space.

In the early days of chemistry, the structure of aromatic compounds was a mystery. The existence of a stable compound consisting of six atoms each of carbon and hydrogen seemed inconsistent with the known properties of atoms. This paradox was resolved in a deep reverie by the German chemist Friedrich August von Kekulé. According to one version of the story, he envisioned atoms that wriggled like snakes, one of which seized its own tail. The image of six carbons forming a ring supplied the key to the structure of aromatic hydrocarbons.

Auditory imagery was a critical component in the compositions of Wolfgang Amadeus Mozart. Segments of melodies would appear spontaneously in his mind's ear, and he would accept or reject them as was appropriate for the musical piece at hand. In a similar way, Henri Poincaré's "sensual imagery" led him to sense a mathematical proof in its entirety "at a glance."

For some physical scientists such as Albert Einstein, imagery is in fact the dominant form of mentation.

> The words of the language, as they are written or spoken, do not seem to play any role in my mechanism of thought. The psychical entities which seem to serve as elements in thought are certain signs and more or less clear images which can be "voluntarily" reproduced and combined.
>
> There is, of course, a certain connection between those elements and relevant logical concepts. It is also clear that the desire to arrive finally at logically connected concepts is the emotional basis of this rather vague play with the above mentioned elements. But taken from a psychological viewpoint, this combinatory play seems to be the essential feature in productive thought. . . .Conventional words or other signs have to be sought for laboriously only in a secondary stage, when the mentioned associative play is sufficiently established and can be reproduced at will.

Some problems naturally suggest images as the means for deriving their solutions. An obvious example is a physical maze, for which a verbal description might be so cumbersome that a potential problem solver would have great difficulty even understanding the situation, let alone reasoning about a solution.

In many other situations, imagery may be more advantageous than verbal reasoning purely for the sake of efficiency. In theory, linguistics and imagery would be equally capable, but in practice the latter may prevail. The advantages of imagery relate to parallelism, acceleration, and multidimensionality.

A key strength of imagery is parallelism. For example, we can monitor the movements and gestures of a half dozen people at a party or on the silver screen. On the other hand, linguistic events are sequential; we cannot usefully keep track of the conversations of these same half dozen individuals as they converse in pairs about unrelated topics.

A second factor relates to the susceptibility to acceleration. In considering visual or auditory images, it is almost as easy to comprehend a picture or a sound at a speed many times faster or slower than the original, referent image. On the other hand, the temporal bottleneck in encoding ideas into long-term memory limits the amount by which linguistic reasoning may be speeded up. We can comprehend a spoken argument presented much more slowly than at a natural tempo, but we lose track of a train of thought presented too quickly.

The third reason for efficiency is multidimensionality. Visual imagery, for instance, is a medium where we can view relationships in two and sometimes three dimensions, thereby providing more "room" for thought.

For reasons of efficiency, imagery is a superior mode of mentation even when linguistic or verbal modes are capable vehicles. It allows us to consider more ideas and thereby make more progress over the same stretch of time.

The importance of visual imagery to creative problem solving suggests the use of computer graphics to enhance the ideation process. For restricted domains of application, this is a relatively simple task. For example, a graphics simulator for creating robotic manipulators might encode all the known designs and geometrics relevant to the problem at hand, plus perhaps a set of decision rules for generating new mechanisms.

A more challenging question is whether a set of tools may be developed that is relatively domain independent. Hence a graphics package to assist in the design of mechanical devices is potentially more useful than one that addresses only robotic manipulators.

Given our current lack of understanding in the areas of creativity, graphics, and artificial intelligence, a two-stage development is likely to unfold. First is the development of a generalized system architecture that may be readily tailored by incorporating domain-specific knowledge such as automotive principles or marketing strategies. Experience with those tools will then suggest certain similarities and may then be exploited to develop a set of domain-independent tools. The nature of these tools remains to be determined, although they will likely bear some resemblance to existing techniques in computer graphics and artificial intelligence.

Existing methods of creativity enhancement tend to focus on words as the key factor for representing ideas. Since imagery is such a critical component of thinking, however, pictures can assist in the initial exploration of ideas as well as in their communication to a larger audience. The generation of prose often helps to clarify our thoughts. In a similar way, the process of preparing drawings or schematic representations—be they of physical or intangible objects—helps to nurture the ideas further. Hence drawing can be an integral part of the idea-generation process: the problem solver expresses her ideas on paper, then evaluates the tentative results for the next iteration.

Such drawings may take the form of pictures, graphs, icons, schematics, block diagrams, flow charts, figures, blueprints, or other graphic representations. In these drawings, words are often mixed in synergistic fashion with pictorial representations.

Externalization Factor

Externalization involves the expression of ideas through concrete or abstract channels. The creation of a sculpture or the construction of a skyscraper involves a tangible expression of a solution. In contrast, singing a song or proving a mathematical theorem involves an intangible form of externalization. Even in these cases, however, a physical medium facilitates the process, whether in the form of air molecules, paper and pencil, or some other vehicle.

The utility of externalization lies in the fact that describing the problem to a second party or committing it to an external medium helps to clarify the ideas and is often an important step toward the solution. One of the most useful vehicles is that of putting ideas to paper. There are several reasons for this:

- Writing down the components of a problem focuses attention on the salient aspects because of the need to assign names, symbols, or pictorial representations to the key objects.

- The limitations of short-term memory constrain the amount of information that can be considered simultaneously. An external medium such as paper or blackboard serves as an extension of working memory.

- The spatial dimension highlights certain relationships among the items of data. For example, tables spotlight missing information, while figures are effective for emphasizing trends in data.

The process of externalization may serve as the vehicle for the further development of an idea. The nature of this effect, as well as procedures for its exploitation, are discussed in Chapter 5 on individual methodologies.

Summary

Creativity involves five factors: purpose, diversity, relationships, imagery, and externalization. The purpose, as a statement of the overall vision, specifies the mission of the creative endeavor. Diversity and relationships are primarily attributes of the solution to a difficult problem, while imagery and externalization are vehicles for enhancing the ideational process.

- The purposive aspect refers to identification of the critical attributes of a problem.

- The diversity factor refers to the association of objects or ideas that might previously have appeared disparate.

- Relationships among the diverse elements of a creative solution can take the form of similarity, contrast, membership, causation, or other connections.

- Ideation through imagery is a critical component of creative problem solving. The document form of ideation—whether visual, auditory, or otherwise—will depend on the individual and her problem domain.

- Externalization of an intermediate result allows the problem solver to gain a fresh perspective and obtain feedback from other individuals, thereby advancing a step closer to the solution. Further, the final solution must take some form of expression if it is to be shared with others.

5

Tactics for Individual Productivity

He felt that he was in a locked room in the middle of a great open country: it was all around him, if he could find the way out, the way clear. The intuition became an obsession. During that autumn and winter he got more and more out of the habit of sleeping. . . . He dreamed vividly, and the dreams were part of his work. . . . He got up and scribbled down, without really waking, the mathematical formula that had been eluding him for days.
                                                    Ursula Kroeber Le Guin

The resolution of a difficult task often follows a series of identifiable steps. Students of creativity have often characterized the process as a four-stage phenomenon:

- *Preparation*: Orientation to the problem and definition of the task.
- *Incubation*: A period of aridity or immersion in unrelated activities.
- *Illumination*: A sudden spark of insight and the recognition of a candidate solution.
- *Resolution*: Assessment of the candidate solution and its implementation.

The preparatory phase of a difficult task involves a careful consideration of the underlying problem and a clear specification of the goals. Too often we tackle tasks without an adequate definition of the true problem, and discover too late that the solution resolves an irrelevant problem.

A distinguishing feature of challenging tasks lies in their resistance to attack, and solutions do not always spring from a single session of wrestling with the problem. Often difficult issues are resolved, whether gradually or suddenly, over spaced efforts interrupted by unrelated activities. Some students of creativity take the view that incubation is a period of "intermission" in which the problem submerges into the subconscious. The lack of strict conventions in this netherworld allows for the juxtaposition of new objects in novel and even bizarre combinations. The resulting ideas possess the novelty that is an essential ingredient of a creative solution.

Other writers take the view that incubation is nothing more than a period of relaxation or recuperation in which a tired mind regains its energies. Still others would claim that the incubation phase is merely a stretch of time that allows the mind to assimilate other stimuli from the environment, whether at the conscious or subconscious level. Whatever the true role of incubation, the fact remains that many of our difficult problems are resolved only after several sessions of conscious effort separated by seemingly unrelated activities.

A candidate solution to a difficult problem often occurs suddenly, just like the elementary ideas that constitute them. Such preliminary solutions or "insights," however, can be misleading or even incorrect. For this reason, the ideas must be properly evaluated. The problem is finally resolved when a candidate solution fulfills its objectives and is expressed in a form comprehensible to other people.

The following attributes have been identified as common elements of creative individuals:

- Capacity for intuitive perception: the recognition of associations and similarities among objects and concepts.
- Concern for implications, meanings, and significances.
- Ability to think imaginatively without regard for practicalities.
- Attitude of open-mindedness and receptiveness to change.

The creative problem solver avoids excessive attachment to similar solutions to past problems. The better approach is to draw on current knowledge and individual experience, rather than rehashing old techniques and outmoded traditions.

An individual can take deliberate steps to enhance his own level of creative problem solving.

Preparation

When an idea is generated, it must be communicated to the external world to be of value to others. The form of communication may be tangible, as the construction of a sculpture or a skyscraper; or it may be intangible, as the creation of a novel or a scientific theory. Other channels of expression, such as a play or a marketing strategy, may exhibit both concrete and abstract features.

Often the very process of communication can refine the creative idea by highlighting its weaknesses or suggesting improvements.

These quality-enhancement practices may be roughly partitioned into three phases: input, processing and output.

Input Phase

Quality thinking requires quality inputs. A child raised in a sparse, plain environment will tend to display limited intelligence—by most formal or informal criteria—compared to one immersed in a rich environment offering a variety of stimuli. In a similar way, a researcher or other creator has to expose himself to a rich spectrum of inputs from colleagues, journals, and other sources.

For difficult problems, it is not clear which way to turn for sources of inspiration. In this case, the only recourse is to skirt the perceived boundaries of the main

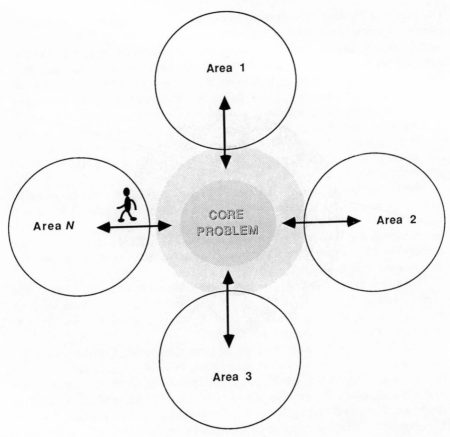

Figure 5.1. Foraging in neighboring areas (1,2,3, . . . *N*) for sources of inspiration.

problem and delve into related fields. Unfortunately, it is often uncertain which of the neighboring areas are the most relevant, or even where the boundaries of the main problem lie. This situation is depicted in Figure 5.1.

To illustrate, suppose that the main task is to develop a science of manufacturing. It is unclear how to determine *a priori* the structure of such a science, or even the boundaries of its jurisdiction. But it is apparent that manufacturing involves the transformation of materials into useful products through intelligent processes implemented at reasonable cost. Hence some referent disciplines to investigate relate to materials science, mechanical engineering, computer systems, and microeconomics, among others.

The case for an interdisciplinary approach to problem solving is supported by the history of science:

Almost always the men who achieve these fundamental inventions of a new paradigm have been either very young or very new to the field whose paradigm they change.

The inertia of existing paradigms and procedures is not limited to scientific activities, of course. Examples are legion in the industrial sector, where novel technologies are brought to fruition by fledgling organizations that then proceed to dominate the marketplace. A case in point is the minicomputer, conceived by a start-up venture called Digital Equipment Corporation. Digital created both the machine and its constituency. The company maintained its leadership in this key market despite fierce competition from IBM and other latecomers.

Studies of innovation in the industrial sector indicate that new products often originate from users rather than vendors or suppliers. In the realm of scientific instruments, for example, 77 percent of innovations spring from users and only 23 percent from manufacturers.

The history of xerography illustrates the tendency for new industries to sally forth from unexpected quarters. Chester Carlson, a patent attorney, invented a method to copy documents on ordinary paper as a way to eliminate the inaccuracy and inconvenience of retyping patent applications. The first patent for this technique was filed in 1937 using the term electrophotography, and the first successful copy was demonstrated the following year.

During the next nine years, Carlson attempted to market his idea to more than twenty firms such as General Electric, Kodak, IBM, and RCA. None of these companies could see any potential for a new contraption whose function could be well served by carbon paper or mimeographs.

Despite the cool reception from the business community, Carlson persevered, demonstrating his technique with manual plates. In 1944 he established a royalty-sharing arrangement with Battelle Memorial Institute of Columbus, Ohio, which would further refine the process. Three years later Haloid, a small company in Rochester, New York, joined Battelle and contributed research funding. The term *xerography*, a neologism based on Greek roots for "dry writing," was introduced to the English language.

The manual copying process involved over three dozen steps, and required the best operators two or three minutes to produce a print. In the thirteen years that followed, Haloid spent more money to improve the process than it earned.

In 1959 the Xerox 914 copier was unveiled. It would be widely recognized as one of the most successful products of all time. It helped to transform the Haloid company of 1959, a $32 million business based largely on photographic paper manufacturing, into the present Xerox corporation, a $14 billion titan with more than 100,000 employees around the world. Over the decades, Xerox has maintained its leadership in the photocopier market in spite of concerted efforts by numerous competitors.

Processing Phase

The processing phase is in some sense the heart of the creative endeavor. Examples of practices in this stage relate to a judicious mix of "foreground" and "background" processing. In the former category, a researcher needs periods of solitude for quiet contemplation. The utility of the second category stems from the fact that the solutions to our toughest problems are often unheralded. These ideas arrive in a "flash," while we are driving, taking a shower or drifting off to sleep. On the other hand, the mind must not be fully engaged in the "foreground" task involving conscious effort. We are not likely to encounter great insights while playing a strenuous game of tennis or struggling with insomnia. In this case, the problem is not in background mode, but suspended altogether.

A number of techniques have been developed to enhance the creative process. The most widely known techniques relate to trigger words, checklists, morphology, brainstorming, and synectics.

Trigger Method. In the trigger word approach, a problem solver asks himself a series of active questions. The trigger word is the verb in the problem statement, a term that refers to the associated function of a product. For example, the function of a car is to *transport* people and goods. The purpose of the vehicle, however, may be viewed in other terms, such as *carry, convey, move, haul,* or *relay*; each of these verbs connotes a variant means of locomotion. The connotations in turn suggest alternative solutions to the transportation problem.

Checklist Method. This approach relies on a number of questions relating to modification or transformation. A set of catalytic words may be used to transform old ideas into fresh ones. For example, the key word *adapt* may conjure up the idea of modifying mass production techniques to the delivery of fast food or to the processing of checks in banks. In a similar way, the key word *substitute* may lead to the displacement of oil by solar panels in heating a house, while the trigger *minify* may suggest the reduction of data storage media from bulky video cassettes into compact optical discs.

Morphological Method. The *morphological* method involves the systematic identification of all relevant options and their combinations. To employ this method, the would-be creator performs the following steps:

- Identify the problem domain.
- List all relevant attributes or dimensions.
- Identify all possible combinations.
- Evaluate each combination.

The morphological approach involves the identification of various parameters and the values they may assume, followed by the consideration of each set of attribute values in turn. The method has been used, for example, to explore the varieties of potential chemical propellants. The situation is depicted in Figure 5.2: chemical energy can take one of two pathways, in any of three physical states, to drive two types of objects through four external media. These numbers imply 48 combinations, any of which defines a potential solution. For example, the list {heat, liquid, vehicle, air} characterizes a solution in the form of an ordinary jet aircraft.

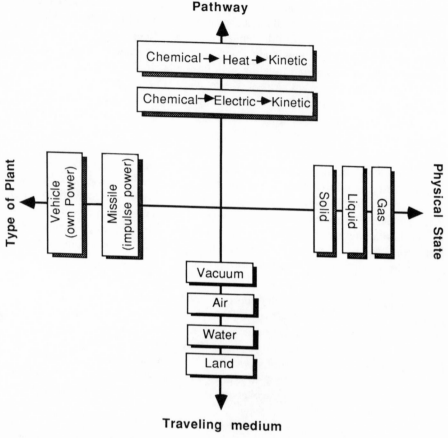

Figure 5.2. Attributes for a morphological evaluation of chemical propellents.

The emphasis in the morphological method is on a complete enumeration and consideration of all potential solutions:

> In all his explorations the morphologist strives for *complete field coverage*. This means that he will not be satisfied until he has found *all the solutions* to a given problem.

Problems of realistic size, however, admit so many potential solutions that computational constraints usually render impossible a complete search of the solution space. Therefore, the morphological approach is appropriate only for problems of limited scope, where the relevant attributes for the problem can be identified in a relatively straightforward way.

The technique is less relevant for difficult tasks. In a corporate environment, a chief executive may be faced with the question, "What steps should our company take to prepare for the 21st century?" Or the robotics researcher may be faced with the question, "What information characteristics are critical to system performance, and what is the nature of this relationship?"

The morphological approach is to analyze the problem and determine the independent parameters involved. These are then listed on a grid and evaluated systematically. A similar approach is the *attribute-seeking* method. The plan here is to single out the basic attributes of a problem, to generate viable alternatives, and to evaluate the resulting candidates.

Brainstorming. The *brainstorming* technique refers to the spontaneous generation of ideas by a diverse group of individuals, some of whom may know little about the particulars of the problem. The object is to generate all possible solutions through free association. Each proposal, however outlandish, is dutifully recorded during the brainstorming session and evaluated only afterwards.

The technique of brainstorming is based on the simple assumption that quantity can lead to quality. According to this view, a large number of candidates will make it probable that at least one of the alternatives resolves the problem.

The methodology involves two stages: *idea generation,* relating to the conception of alternatives, and *idea evaluation,* relating to the assessment of their utility. The first stage may in turn be partitioned into *fact-finding,* in which the purpose of the creative effort is identified, and *idea-finding,* in which existing alternatives are combined, modified, or extended.

The brainstorming procedure is based on two premises:

- Creative people have tolerant test criteria, while uncreative people are too stringent in the filtering stage.
- A "generate-and-test" procedure is effective for the development of ideas. The first phase involves the spawning of concepts, and the second phase their testing or filtering.

The operational implication, then, is to relax the filtering mechanism. The basic premises appear reasonable, and brainstorming has become a standard term in the daily lexicon. In a brainstorming session, participants gather in a group and freely express ideas that are recorded without discussion or criticism. This phase is free-wheeling: each participant is urged to call out anything that comes to mind, without regard for its feasibility or utility, and only minor concern for its very relevance to the subject at hand. In the second stage, the ideas recorded in the first stage are evaluated systematically.

Brainstorming is often helpful for well-defined problems that can be readily recognized by all the participants. It has been used to great advantage in many domains, especially in business applications. Examples are found in the generation of advertising slogans or ideas for new household products. In these situations, novelty is in itself an important ingredient of the solution. For such cases, one candidate solution need not bear any rational relationship to the previous or the next idea.

The technique, however, has its limitations. The methodology would be effective if the cost of generating ideas were insignificant, and no time constraints existed for resolving a particular task. But where do the ideas come from? They spring from thoughtful individuals whose time is not free. Suppose, for a moment, that the procedure for generating ideas at random *were* fully automated. Would it then be reasonable to generate alternatives at random? A clear result from research in artificial intelligence is that complete search is not a viable procedure even for restricted domains such as chess, let alone problems of realistic complexity.

Quantity in idea generation might be helpful when a decision maker is clueless about the solution of a problem. However, sheer quantity is not a sufficient condition for success. Even if it were sufficient, it is not practical in human or computational terms; and further, it is not always a necessary condition.

In a word, the main liability of brainstorming is inefficiency. The fact that 150 ideas, rather than five, are generated in one hour is of little significance if 99 percent of them are useless or irrelevant. The metric of importance is the utility of the ideas generated in a given time, not their number alone. Many psychological studies show that the total number of good ideas produced through brainstorming does not differ from that of nonbrainstorming sessions. In fact, in some instances, the average quality of ideas can be higher *without* brainstorming.

Brainstorming can lead to the whole being less than the sum of the parts. The union of alternative solutions obtained from individuals working alone is often superior both in number and quality to the output from a single group. The real value of brainstorming seems to lie not in the generation of ideas, but in the promotion of consensus which is so vital to the subsequent implementation of real-world decisions in the face of viscous organizational processes.

Brainstorming in its orthodox form may be helpful when the participants are completely unertain about the appropriate approaches to pursue. Even in this case, a short brainstorming session is better terminated by a consolidation phase: a systematic evaluation of the proposed ideas, and some tentative conclusions. These conclusions may then guide the participants for the next phase, which may be more focused discussion or even "directed brainstorming."

If productivity is defined as the total utility of ideas generated per unit of time, then the quest for a creative solution must be guided. To proceed otherwise is to ignore the additional information unearthed during the process of digging for a solution.

Synectics. The synectic technique is another method for use by small groups of individuals having diverse backgrounds. It is similar to brainstorming, but utilizes a more structured approach based on the following two principles.

- Make the strange familiar. The first step in the problem-solving process is to understand the problem.

- Make the familiar strange. The active stage of this technique is to view familiar objects in a new light through analogy.

Four types of analogy are used to make the familiar strange.

- *Personal analogy or identification.* The problem solver identifies with, and imagines herself as, the object in question. If the designer of a lightweight engine imagines herself as such a motor, she may gain new insight into the problems of generating power, transmitting torque, maintaining reliability, and other issues. These critical parameters in turn may motivate novel solutions.

- *Direct analogy or transference.* This method refers to the direct comparison of similar facts or situations. Knowledge and techniques from one domain are utilized in another.

- *Symbolic analogy.* The problem is regarded through objective, impartial images. Verbal descriptions of a problem are replaced by pictorial representations, mathematical formulas, or other media.

- *Idealization or fantasizing analogy.* The problem solver identifies the characteristics of the ideal solution, then seeks ways to realize the wish.

The synectic approach attempts to identify the fundamental concepts underlying a given situation, rather than emphasize the obvious characteristics. This technique offers a much broader view of a problem, and is especially useful when radical departures from traditional concepts are required.

The techniques described in this section were developed in the context of practical applications. They have been employed successfully, for example, to generate ideas for new products that might appeal to an existing base of customers. These applications have limited scope and are amenable to precise definition, as exemplified by the goal of inventing a better can opener. Such freewheeling approaches, however, are inappropriate in many complex situations such as basic research. In these arenas, the number of potential solutions increases exponentially with the number of parameters, but many of the resulting alternatives are worthless. Hence more systematic approaches are required for applications in research and development, as well as for the multitude of problems arising in our daily lives. These issues are considered further in Chapter 6 on research strategies.

Output Phase

The output phase refers to the stage of recording ideas, whether full-fledged or half-baked. The development of ideas is often a boot-strapping affair. Sometimes the complete solution to a problem appears suddenly from unknown quarters, even as we mull over it on paper with no more intent than to clarify the problem statement, or perhaps while we explain it to a friend.

At other times, we may be writing down a partial solution or preliminary set of results, and new ideas occur as we write. For example, we may discover deficiencies and opportunities in the argument while reading a preliminary draft, thereby paving the road to the next draft.

Interphase

Certain practices for problem solving apply to all three phases above. Per' most common example is that of a supportive circle of colleagues. Henc' productive team is not one that splits up at 5 p.m. each workday, but ' meets for dinner or picnics together on the weekend.

A related concept is that of the critical mass needed for creative problem solving. For small problems, a clever individual can single-handedly arrive at an appropriate solution. When the requisite level of effort is higher, a single person may initially develop some ideas but will soon reach a dead end.

The value of group interaction to stimulate creativity is well known. For large problems such as the development of a theory of manufacturing systems, synergism is required among colleagues in a community of shared interests. One person may generate an idea that is refuted or refined by someone else, then developed further by a third or even the originator himself, and so on.

Method of Directed Refinement

Most existing techniques for enhancing creativity deal with the straightforward procedure for generating all possible combinations. The goal of brainstorming, for example, is to generate a barrage of potential solutions, regardless of how reasonable or bizarre they may be, in the hope that some of them will withstand the test of reason. This may be useful for developing a new gadget to mow the lawn or paint fingernails, but it is less helpful for composing a poem or developing a science of intelligent systems.

In addressing difficult problems, a more focused approach is needed to manage the exponential explosion in alternative pathways. In the realm of research, for example, how does one go about developing a theory of intelligence? Does one look at applications first, or attempt to develop a preliminary theory? Should biological systems be considered first, or artificial devices, or both concurrently? Does one develop a hierarchy of concepts, or generate a network or some other form of organization? Should continuous phenomena or discrete characteristics be modeled first? Are continuous mathematical tools such as calculus and differential equations appropriate, or discrete techniques such as logic and computability theory? There is almost no end to the different combinations of approaches that may be pursued, some of which may be helpful and others not. Further, any particular combination of tactics and methodologies could easily consume a lifetime of research.

For such large, open-ended problems, a strategy of progressive refinement is more appropriate than blind search. More specifically, the appropriate procedure is one of directed refinement consisting of the following stages:

1. *Breadth-first search.* A preliminary investigation of potential tactics and techniques provides a better orientation to the problem and an "intuition" for the most promising avenues. The global scan includes a consideration of new results and potential developments in the external environment, as well as an assessment of personal capabilities and organizational resources.
2. *Depth-first search of the identified path.* The most promising approach identified in the last phase is pursued until (a) some results are obtained, or (b) the determination is made that further search along that path is not worthwhile.

3. *Evaluation.* The work to date is evaluated to determine whether (a) the problem has been solved, (b) the current path should be explored further, (c) a new path should be pursued, or (d) the project should be abandoned.

4. *Decision.* If the problem is solved or is about to require excessive further effort, then stop. Otherwise, determine whether a breadth-first search of a new subset of topics is appropriate, or a depth-first investigation of some particular path; then go to Step 1 or Step 2, respectively.

The Directed Refinement method is a composite of various problem solving strategies as they relate to difficult tasks. The technique is appropriate not only for research, but for difficult problems in other realms as well. For example, Picasso pursued many different avenues for expressing his condemnation of war, as reflected in the numerous preliminary sketches for what would become one of his greatest works, the painting *Guernica*. In this effort, it is apparent that Picasso drew on a repertoire of previous works, by both himself and others (breadth-first search); he then generated many preliminary sketches (depth-first search), rejecting one after another (evaluation and decision) until he finally settled on a satisfactory design (decision).

The method of progressive refinement is consistent with the behavior of skilled problem solvers in any walk of life. The psychologist Robert Sternberg asserts that the allocation of resources, a task he calls *metaplanning*, may be the most prominent aspect of effective problem solving. This activity is facilitated by four strategies.

- *Willingness to allocate a large portion of the available time to high-level planning.* Most people spend too much time delving into minutiae that are of little or no value; a measured consideration of global issues would highlight the relative importance of different topics. A simple example is a test-taker who seeks the answer to an imaginary problem because he did not read the instructions properly.

- *Making full use of prior knowledge in planning and allocating resources.* In performing a set of errands, an efficient path can be developed by visualizing the geographic dispersion of the locations and their interconnecting links. In a technical domain, if geometric and algebraic techniques seem equally promising for solving a problem, then an investigator possessing greater skill in algebra should pick that technique first.

- *Flexibility and willingness to change the plan.* Pursuing a plan may highlight difficulties which suggest a modified approach. A student who has opted to write a paper on the mating habits of kangaroos would be wise to select a different topic if he cannot obtain enough background material. A marketer who finds that her product does not sell through magazine advertising might try radio spots.

- *Watching for new kinds of resources.* Often people treat real-life problems like homework assignments: the problem is defined at the outset for all time, as are the tools. In the larger game of life, the available resources change over time, for better or worse. The design of a lightweight engine may be facilitated by the discovery of a new alloy; on the other hand, the problem may be hampered by the legislation of stricter standards on automotive emissions.

The first heuristic relates to formulation of strategy. This includes a proper orientation to the problem, a global perspective of the issues, an overall plan of attack, and a tentative agenda. The second item calls for the full use of available resources, including previous know-how and experience. The third heuristic relates to the dynamic, on-line monitoring of progress on the problem. The information should be used to determine how and when new developments should be factored into a reformulation of the plan of attack. The last heuristic relates to the monitoring of events external to the basic problem. It involves keeping track of novel developments that may expedite the quest for a solution, or even enable the resolution of an otherwise insolvable problem.

Principle of Generality

The nature of a difficult problem may change over time. This happens, for example, when the underlying problem metamorphoses. Consider the manufacturer whose raison d'être was to serve society with high-quality, low-cost slide rules. His mission was annulled by the advent of the electronic calculator.

More often, the potential solutions may change over time. This is especially true when the problem-solving task takes a long time or the relevant technologies develop rapidly. An old movie recounts the story of an astronaut who takes off on a starship amid great fanfare, the first human to attempt to reach a star outside our solar system. His ship, traveling at a fraction of the speed of light, completes the journey safely and eventually returns to earth. In the meantime, a century has passed on earth, but the traveler has aged only a few years due to relativistic effects. He steps out of the ship, expecting to be greeted with a hero's welcome. But only his great-granddaughter has come to receive him. The rest of the world had long since forgotten the would-be hero: later ships, traveling even faster than his, had completed the journey years before he did.

We do not routinely encounter space travelers, nor a century as a scope for planning. But the story is particularly apt in today's world, characterized as it is by fast-paced sectors such as the computer industries, where three years may separate one generation from the next.

We cannot predict the future, especially when addressing difficult problems. But a relevant principle of action when trying to find enduring solutions to problems is the following.

Principle of Generality. Other things being equal, pursue general
or enduring topics.

In other words, delve into the topics that pertain to several alternative futures, rather than only one.

This dictum is consistent with the view of formal education as a vehicle for imparting to students general principles rather than disjointed facts. Students should be exposed to the diversity of knowledge and the skills of learning how to learn rather than trained solely to recite facts and practice specialized techniques. These generalized capabilities will, ideally, provide the student with a broad and robust base from which to pursue personal and professional interests.

In conducting research, an effective strategy is to begin with topics that will apply to two or more pathways to the solution, rather than just one. For designing a more efficient propeller, it is better to investigate the thrust and viscosity of different shapes, rather than the reliability of a particular material of manufacture. The aerodynamic characteristics will likely apply to a variety of materials; but reliability considerations may be rendered irrelevant by a new material that is invented in the intervening years.

An implication for product innovation is the maintenance of an open architecture, allowing for the incremental enhancement of an initial product configuration. Until the early 1980s, machines from the Apple Computer company incorporated a closed architecture: stand-alone devices without the capacity to communicate with machines from other vendors. Even a user who needed a more sophisticated computer from Apple was expected to discard his old machine, much like an obsolete appliance. Despite its early lead in the personal computer market, Apple suffered heavy losses in market share to newcomers such as IBM, which offered an open configuration. In 1984 Apple Computer's very survival seemed questionable. Thereafter the company opted for coexistence with other vendors, adopting an open configuration and regaining its prominence in the hardware wars.

The moral for other innovators is to seek endurance through generality. For long-term viability, a developer of a factory automation system must maintain an open architecture to accommodate the new developments occurring both in-house and externally.

In the business world, generality of function does not necessarily mean generality of markets. The increasing expectations of personalization from consumers have led to market segmentation. With increasing frequency, the most successful firms are those with superior products designed for particular market niches, rather than mass marketers offering commodity products. The reward for specialization is premium prices. In this context, a vendor of factory automation systems will possess a core set of machines, techniques, and know-how that cut across different industries. Rather than offering a single all-purpose factory automation system, the firm will more likely succeed with a portfolio of packages for specific industries, whether for the manufacture of cars, socks, or lipsticks.

Summary

The creative process often requires four steps: preparation, incubation, illumination, and resolution. Each of these steps may be enhanced through the deliberate application of general strategies and specialized tactics.

- Most creative individuals exhibit the following characteristics: capacity for recognizing associations and similarities among objects and concepts; concern for implications, meanings, and significances; ability to think imaginatively without regard for practicalities; and an attitude of open-mindedness and receptivity to improvement.

- An individual can take deliberate steps to enhance personal creativity.

- The quality of output depends on the quality of input and of the intermediate processing activities.

- Interphase refers to the germination stage of ideas. It is often promoted through interaction among a supportive circle of colleagues and other interested individuals.

- The processing phase can be enhanced through a number of techniques such as the *trigger word* method, the *checklist* technique, the *morphological* method, the *brainstorming* approach, and the *synectic* technique. However, these straightforward approaches tend to generate candidate ideas wantonly, many of which are of low quality.

- The output phase of the creative process involves the expression or recording of ideas.

- For difficult tasks, the Method of Directed Refinement is an effective procedure for exploring the problem domain and generating candidate solutions.

- A guiding rule for addressing difficult problems is the Principle of Generality: Other things being equal, pursue general or enduring topics.

6

Managing the Project

Science seldom proceeds in the straightforward logical manner imagined by outsiders. Instead, its steps forward (and sometimes backward) are often very human events in which personalities and cultural traditions play major roles. James Dewey Watson

How does one approach a large, difficult problem? By definition the solution to the problem is ill-defined, and the path to the solution is even more obscure.

The poor definition of the problem does not, however, imply the complete lack of operational strategies. It is not sufficient to throw up our hands in despair and go fishing to await divine inspiration.

The resolution of difficult problems—and of easy ones—can be facilitated by a coherent strategy. This "rational" nature of problem solving applies to the realm of scientific discovery:

> However romantic and heroic we find the moment of discovery, we cannot believe either that the events leading up to that moment are entirely random and chaotic. . . . We believe that finding order in the world must itself be a process impregnated with purpose and reason. We believe that the process of discovery can be described and modeled, and that there are better and worse routes to discovery—more and less efficient paths.

Purposive activity enhances creative problem solving not only in science, but in nontechnical arenas as well. This chapter discusses a number of strategic issues and techniques for addressing challenging problems.

Types of Failure

Some failures are productive while others are not. An *active* failure is one that serves to advance the state of knowledge. Such a result may be used to modify a tentative hypothesis, whether through refinement or outright rejection. Active failures may in turn be classified into two types: definitive or mixed.

A *definitive* failure is a strong result that may be used to overturn a proposition by showing it to be false. To illustrate, consider the question, "Are there stars in our galaxy that are over 15 billion years old?" Suppose that a means were found to give a definitive answer, and the resulting answer was "No." This type of failure is

51

actually a positive result for the opposite hypothesis. In other words, the negative response is actually a positive confirmation of the query, "Are there *no* stars in our galaxy over 15 billion years old?" In fact, a statistician might well have begun the investigation with the null hypothesis, "No star in our galaxy is over 15 billion years old."

A *mixed* failure is a weak form that is part victory and part defeat. This type of result serves to modify the germinating hypothesis. To illustrate, consider the query, "Is information theory relevant to the development of a science of manufacturing?" Suppose that the response is, "The basic concepts and formalisms are relevant, but these must be attributed with a new interpretation or semantic perspective." The original query is answered partly in the affirmative, and partly in the negative. Hence this is a weak form of active failure that is itself a positive result. It may be used as a stepping stone to further progress.

Both categories of active failure are actually successes in disguise. The mixed failure may perhaps be the most common type of result in addressing a difficult problem, especially in the intermediate stages of developing a solution. Unfortunately, inexperienced problem solvers sometimes confuse the hybrid form of failure/success with a completely different result: the passive failure.

The *passive* failure refers to the lack of results, whether positive or negative. In the manufacturing science example, an investigator might pursue topics such as information theory or mathematical logic with the hope of adapting them to the realm of industrial production. However, he makes only a half-hearted effort to understand the referent disciplines and gives up before he is in a position to decide whether or not they are relevant, or to ascertain the ways in which they are not. The only "conclusion" he can draw is, "I don't see how these topics are relevant."

Had the investigator delved more deeply and ascertained the specific drawbacks and limitations of, say, symbolic logic for formalizing manufacturing knowledge, then he would have attained a mixed success/failure. He might then have been able to build on his preliminary results and adapt or extend the referent disciplines to his problem domain.

Difficult problems, by their very nature, do not admit straightforward solutions. Hence a categorical result—whether in terms of an unalloyed success or its converse, the definitive failure—is a rare phenomenon. The novice problem solver, however, often works under the "all-or-nothing" principle. In the completed work of others, the novice sees only the end result of a painting, the printed version of a novel, or the final structure of a mathematical proof. This perspective is unfortunately reinforced by the way in which the product of creative work is presented. It is a rare article or textbook that explains all the false starts and revisions during the creation of a poem, symphony, or theory. To spare the observer the frustration and anguish of the development process, the authors of these objects almost always present them as beautifully sculptured and polished pieces, as if they were the product of a straightforward, linear process. But in reality this is seldom the case. Research is a more chaotic, jumbled exercise than the sleek process suggested by published accounts. So the inexperienced problem solver merely frets, wrings his hands, and waits in vain for the great inspiration that never comes.

Even in the sciences, advances occur through trial and error, and knowledge accumulates in an opportunistic fashion. In the words of the mathematician George Polya:

> Many mathematical results were found by induction first and proved later. Mathematics presented with rigor is a systematic deductive science but mathematics in the making is an experimental inductive science.

The experienced problem solver thrives on the partial results of mixed failure, for they are usually the only kind of intermediate results available. These results are fashioned into a finished product through a process of gradual refinement.

This point is highlighted by Frederick Brooks, a computer scientist who managed the development of the operating system for the IBM 360 family of computers.

> Where a new system concept or new technology is used, one has to build a system to throw away, for even the best planning is not so omniscient as to get it right the first time. . . . Hence plan *to throw one away; you will anyhow.*

Problem solving is a process of selective trial and error. The harder a problem, the greater the number of false starts and intermediate plateaus.

The recognition of the difference between active and passive failure is a key transition point for novice researchers such as beginning graduate students. The realization that progress occurs as a series of false moves into blind alleys—two steps forward and one step back—sets the stage for systematic progress.

Thomas Alva Edison claimed that genius is 1 percent inspiration and 99 percent perspiration. A longitudinal study of over 1000 gifted individuals covering a period of 3 decades has resulted in a similar conclusion.

> Personality factors are extremely important determiners of achievement. The correlation between success and such variables as mental health, emotional stability, and social adjustment is consistently positive rather than negative. . . . The greatest contrast in the two groups [the most and least successful] was in all-around emotional and social adjustment, and in drive to achieve.

Hence success in solving specific problems, as well as in life-long achievement, depends more on focused effort than providence or innate ability.

Strategies

Milestones

One way to speed up the drive to the goal is to seek a deliberate series of intermediate results to serve as stepping stones. By examining the strengths and limitations of an intermediate result, it is often possible to discern a promising direction for the subsequent search. If the result is a total failure, one has even then learned which paths are fruitless. Often failure can be more instructive than success, if not as emotionally satisfying.

In the early days of the U.S. space program, the rockets disintegrated on the launch pad, sought earth-bound targets, or otherwise misbehaved. Even so, the observation was made that each failure contributed more to space technology than did any corresponding success. A failed approach spotlights gaps in one's understanding; a success merely validates what was already presumed to be true.

To accelerate the movement toward the final goal, it is necessary to take risks. Hence an effective policy is the following rule:

Principle of Accelerated Failure. When the cost of failure is low,
fail quickly and often.

This policy may not be so wise when the cost of failure is high, such as when human life is at stake. But it is entirely appropriate when the cost is low, as in the development of theoretical models for intelligent systems.

The psychological impact of even tentative results may be more important than their actual value in generating the final solution. Nothing motivates like success, even if it is partial or tentative.

At first glance, the notion of accelerated failure might appear to be a close cousin of brainstorming. On second thought, however, there are important differences. In spirit, brainstorming is a freewheeling activity in which ideas are generated in rapid succession, without taking time to evaluate the utility of any concept or determine the relationship among the potential solutions. In accelerated failure, on the other hand, the results of one trial are critical for determining the next.

A second difference in the two approaches lies in implementation. In brainstorming, the ideas are generated by several participants in a single session. Accelerated failure implies the testing of ideas over time, by one or more individuals, through a sequence of cycles: pursuing one idea, evaluating the outcome, and using the results as the basis for the subsequent approach.

Studies of creativity in both science and art support the hypothesis that the likelihood of obtaining successful results does not vary significantly from one individual to another, nor among projects by a single individual. Rather, the number of successes depends on the number of attempts that are made:

> Perhaps the odds that any single contribution will prove successful are constant across all creators and so those creators who are most likely to produce a masterpiece are precisely those who produce more works altogether.

In all fields of endeavor, the most eminent individuals also tend to be the most prolific. Since the prior likelihood of success of one thoughtful attempt is as high as any other, the number of trials is a key correlate of success. This was certainly the case in Thomas Edison's tenacious efforts to discover a useful filament for the electric light bulb.

Internal Deadlines

In problems of any size, time is a limited resource. Even so, procrastination seems to be a problem-solving "technique" favored by each of us. When there are things to be done, we tend to put them off till the last day or minute.

Perhaps procrastination is an instinctive response to the fact that we cannot compress our lives into a single day. Why work today when you can put it off till tomorrow? Sometimes problems disappear of their own accord over the course of time. On the other hand, the opposite view has much to recommend itself as well, for "a stitch in time saves nine."

The pervasive problem of breaching deadlines when pursuing difficult problems might be called the "80/80" rule: the first 80 percent of the work takes 80 percent of the time, and the last 20 percent takes another 80 percent.

The cost of a project depends on the product of the number of workers and their workdays. Unfortunately, progress does not yield to such a simple formula.

> *The man-month as a unit for measuring the size of a job is a dangerous and deceptive myth.* It implies that man and months are interchangeable.

Workers and days are tradeable only for simple tasks that may be partitioned into relatively independent subtasks. This is not the case for a complex task which has many interdependent components, let alone those that present conceptual challenges. The relationships between progress and the input effort are shown for different types of tasks in Figure 6.1. The figure indicates, for instance, that calendar time can be compressed by increasing the number of workers, if a project can be broken down into subtasks with little or no intercoupling. On the other hand, a

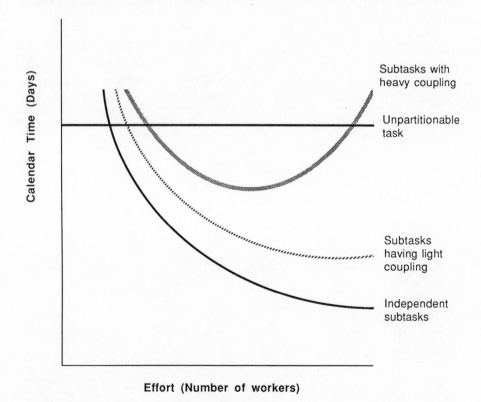

Figure 6.1. Calendar time required to complete a task is a function of the input effort and nature of subtasks. [After Brooks (1975), pp. 16–19.]

project with many dependent or coupled modules may suffer when too many cooks contribute to the broth.

The field of operations research focuses on the identification of optimum solutions in the face of various constraints. If time is a resource, then its availability is a constraint on the problem-solving process.

When the constraints are tight, the optimal solution will entail greater overall cost, or will be inferior to the case where the constraints are lax. An obvious corollary is that the overall effort required to solve a problem can only increase, or at best remain the same, when there are five days left as opposed to ten. In other words, procrastination tends to increase the cost of obtaining results.

Often the cost increases entirely out of proportion to the amount of time *un*available for completion. This relationship is depicted in Figure 6.2. For example, when the deadline is a day away rather than a week, we must mail a package by special courier rather than by regular mail; or we may have to forego a mailing completely and make do with only a telephone call. We can also be sure that *that* will be the day when the computer system crashes, or the electricity fails, or the couriers go on strike, or a snowstorm closes the office.

And suppose that the deadline is, in fact, met. What is the attendant toll in physical and emotional terms when the project requires another 46 hours to meet a deadline that lurks only 48 hours away?

Yet too often we subscribe to the philosophy, "Delay and maybe the problem will vanish" rather than, "Do it now, for the sake of cost *and* quality." How do we reconcile the difference between what should be done and what we would rather do?

One strategy that we all know and usually ignore is to prioritize. List the tasks by priority; work on the simple ones first. When the task is gargantuan or difficult,

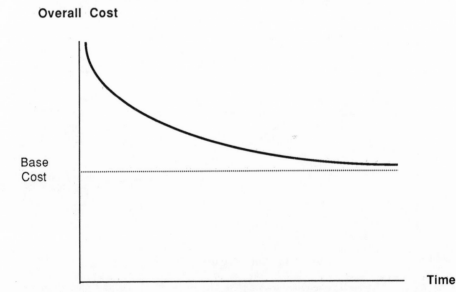

Figure 6.2. Overall cost of solving a problem as a function of the available time.

we are intimidated even before we begin. An appropriate tactic then is to interweave it with other activities, spending only a few minutes on the problem at a time.

One of the *Star Trek* movies features a scene in which the captain and his chief engineer are discussing the time required for repairs to their starship. The scene proceeds roughly as follows:

"How long will it take to repair damages?" asks Captain Kirk.
"About four weeks," says Engineer Scott.
"Four weeks!" protests the Captain.
"Well—if you're really in a hurry, I can do it in one week."
"Scotty, is it your custom to inflate repair estimates by a factor of four?"
"Aye, Captain. How else can I maintain my reputation as a miracle worker?"

This interchange exemplifies a strategy for meeting deadlines. The procedure is to identify each high-priority item, determine the external deadline as some number of days T, and set an internal deadline of T divided by 2. If the project is not completed by the internal deadline, there is still time to finish and even polish it. If it *is* completed through some divine combination of genius and luck, then the lucky participants can go fishing. Or even start on the next project.

From these observations we may infer the following heuristic:

Early Bird Principle. To obtain a solution at reasonable cost, set an internal deadline well before an external one. Or, equivalently, accept no external deadlines without enormous slack.

It's better to deliver too early than too late.

Breadth and Depth

How does one approach the search for a solution? Is it better to perform a breadth-first or a depth-first search for the solution? The breadth-first method involves a systematic search among alternative approaches, one layer at a time. The depth-first procedure involves pursuing the most promising avenues first, following a line of investigation until the goal is attained or failure discerned. If the approach fails, the second best approach is attempted in turn, and so on. An analogy may be made to the realm of mining. The breadth-first approach is reminiscent of strip mining, in which earth is removed one horizontal layer at a time. The depth-first strategy corresponds to shaft digging, where tunnels are dug in narrow columns deep into the earth's crust. The distinction between these two approaches is portrayed graphically in Figure 6.3.

The issue of breadth versus depth arises in diverse application domains. In the field of materials science, how does one seek a substance that exhibits superconductivity at elevated temperatures? Does one try all reasonable combinations of substances, or only the most promising ones in turn?

In the field of production engineering, suppose that information theory seems like a promising referent discipline to incorporate into a theoretical foundation for manufacturing science. Should the investigator attempt to absorb the entire field of

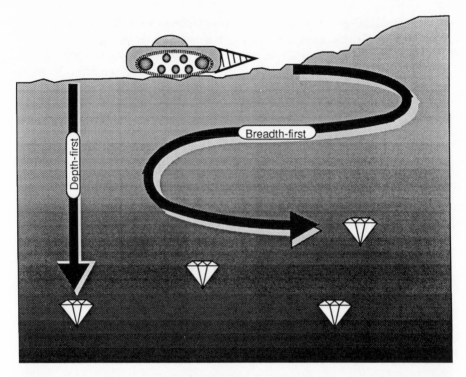

Figure 6.3. Breadth-first vs. depth-first search. The most efficient search strategy depends on the expected distribution of nuggets. Often the best strategy involves the synthesis of both approaches.

information theory before applying its concepts and techniques? That could take years of study—and with continuing progress in the field, it may well be an unending travail. Or should one skim the field and attempt to apply merely the basic principles? That could result in a distorted perspective; often too little knowledge is more dangerous than none at all.

In these circumstances, as in many other hard problems, the best strategy may be one of progressive refinement: perform a breadth-first search, attempt to apply the lessons learned, then use the intermediate results to develop a more informed search strategy.

Similarly, how does one approach a direct mail campaign in the field of marketing? Is it preferable to systematically test all reasonable combinations of brochures, prices, and terms? Or does the marketer select only a few?

The best choice will depend on the specifics of the problem. For example, how likely are the "promising" approaches? If a few seem significantly more promising than the others, then a depth-first strategy for this special set would be appropriate.

Another situational factor is that of time. If a problem has a tight deadline, then perhaps a depth-first policy is preferable only because the systematic approach will not yield a solution in time.

An important characteristic of breadth-first search relates to the mapping of the limits of a new domain of inquiry. In exploring a conceptual space, a useful strategy is to identify its major geological formations before attempting to chart each river, hill, and tree. The observation is summarized in the following postulate.

> **Breadth-First Principle.** In exploring a new conceptual space, determine the major internal features and identify the boundaries before delving into minutiae.

This heuristic is especially appropriate when the field of investigation is so new that a series of depth-first searches would be based on randomness rather than informed guesswork. The breadth-versus-depth trade-off occurs not only in searching for new knowledge, but in the partitioning of efforts between gathering new information and examining the implications of existing information.

Sometimes the general texture of a conceptual space may be known to a limited degree. Even so, the relative worth of alternative approaches might be unclear. In such cases the best strategy may be to pursue both tactics in parallel. In this way, the results of depth-first and breadth-first search can complement each other:

- A breadth-first search will yield quick results, even if they are only of the negative kind. Both successes and failures will suggest promising directions for subsequent search, as discussed previously.

- The intermediate results of a systematic search will provide insight into the nature of the problem, and thereby suggest promising directions for subsequent depth-first search.

By employing these two tactics in parallel, the forays into depth-first search will become increasingly focused and effective. In this way the final solution will likely appear during one of the later depth-first forays following a broad sweep.

Parallel Processing

We usually work on several projects concurrently: our mental facilities are occupied at one moment by one task, then switched to a second task, then back to the first. There seems to be some optimal level of parallelism: too few, and we do not make the best use of our time; too many, and we produce little or nothing.

For example, suppose we are about to embark on a new project, whether it is building a dancing robot or devising a corporate strategy. We make calls to set up appointments with relevant individuals or order equipment from a catalog. If there is no other project in our sphere, then the time interval until the meeting or the arrival of the parts may not be usefully spent. Further, we cannot effectively concentrate on a single project or problem eight hours straight, day after day. So we diversify and pursue more than one project at a time.

The value of pursuing several possibilities in parallel is as applicable to product innovation as to scientific research. In the words of a champion of intraorganizational entrepreneurship:

> To check out one idea at a time has several disadvantages. First, because you are constantly receiving random information from what you read and from people you

talk to, having a number of back-burner ideas gives you a greater likelihood of finding uses for information you pick up. Second . . . it is very hard to be objective when you are down to your last idea.

On the other hand, if we pursue too many projects concurrently, overall efficiency will decrease—perhaps even to the point of zero results. There are too many administrative or secondary aspects to monitor: Where did we put the ninth report from the 37th task force? Is this the motor for the dancing robot or the skiing penguin? And so on.

The concave behavior of overall efficiency as a function of parallelism is also exhibited by computer systems. A time-shared computer system loads a program into main memory, works on it for some quantum of time, then switches it out to bring in a new problem. But the input and output operations themselves take a nontrivial amount of time. When many programs are being executed concurrently, the computer may spend more time on housekeeping duties than on executing the application software. The situation can even lead to "thrashing," a mode in which all the computational energy is spent switching from one program to another, leaving none for actual work. The overhead for pursuing tasks in parallel has similar consequences for human and machine. There is an optimal number of projects that may be pursued concurrently, as depicted in Figure 6.4.

The existence of an optimal level of concurrency is related to the infamous *Principle of Prioritization*, which calls for the identification of tasks and their resolution in order of priority. According to the rule of concurrency, we should prioritize the tasks at hand, identify the top few as the current set of active projects,

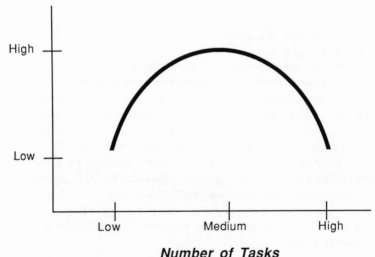

Overall Efficiency

Figure 6.4. Overall efficiency as a function of the number of concurrent activity streams.

and pursue these in parallel. The actual number—say two or eight—will of course depend on the nature of the tasks and the associated overhead. A prescriptive corollary of the prioritization postulate is the following rule:

Principle of Optimal Concurrency. Pursue a moderate number of high-priority tasks in parallel. Do not commit to too many or too few.

A moderate level of parallelism is effective for dealing with difficult problems. In the extremes, over-stretching can be as unproductive as under-reaching in the deployment of resources and capabilities.

Another advantage of focusing on high-priority items is that some problems vanish over time of their own accord. This happens, for example, when a nasty collection letter is followed the next day by a receipt for payment of your bill.

On the other hand, most problems do not disappear over time, and some even become more aggravated. This occurs when a minor report is in danger of missing its deadline, or your car makes strange noises, or your beloved complains of lack of attention. Problems of this type seldom resolve themselves; and putting them off can exacerbate the difficulty or, at minimum, act as sponges on your limited mental energies. No *one* of these secondary problems will stunt your concentration, but three dozen of them will collectively break your stride. Some people suffer death by a thousand nicks in this manner. Our chain of thought brings us in line with Henry David Thoreau's philosophy of the joys of leading a simple life.

Principle of Simplicity. Keep life simple: minimize commitments of secondary consequence.

Despite all efforts to the contrary, however, small problems will crop up with unwelcome frequency. The collection of low-priority tasks that can be dispensed with a few minutes of attention should be thrown into a bundle called "Shorts." This category should be included as a compound project in the active circle of high-priority tasks, and pursued in parallel with the other current projects. We summarize this as a corollary of the Simplicity Principle:

Corollary. Gather brief tasks into a category called Shorts, and work on this compound project in parallel with other high-priority items.

Each task to be addressed, no matter how small, bears a penalty or mental overhead. Eliminating these short items brings no great victory, but it frees the mind for more important tasks. Further, pursuing these secondary items in parallel with other high-priority problems ensures that important projects are not neglected.

Summary

The strategic management of a difficult project requires an effective balance of action and reflection. Too much analysis without sufficient exploration or experimentation leads to stagnation, just as precipitate action without deliberation leads to dissipation.

- An active failure is one that serves to advance the state of knowledge. It may fall into one of two categories: definitive failures and mixed failures.

- A definitive failure is a strong result that may be used to disprove a proposition or, equivalently, validate its counterstatement.

- A mixed failure is a weak result that may be used to modify a germinating hypothesis.

- Any active failure is therefore a success in disguise.

- Seeking a deliberate series of intermediate results, or milestones, may accelerate progress toward the final goal.

- To obtain a solution at reasonable cost, set an internal deadline well before an external one. Or, equivalently, accept no external deadlines without enormous slack time.

- A useful strategy for exploring a problem domain lies in the Breadth-First Principle: In exploring a new conceptual space, determine the major internal features and identify the boundaries before delving into minutiae.

- Breadth-first search may yield quick returns, even if they are only of the negative kind. Both successes and failures will suggest promising directions for subsequent search.

- The intermediate results of a systematic search will provide insight into the nature of the problem, and thereby suggest promising directions for subsequent depth-first search.

7

Strategies for Research

Nature plays fair and if, after climbing one range of mountains, the physicist sees another on the horizon before him, it has not been deliberately put there to frustrate the effort he has already made.   Norbert Wiener

The purpose of research is to acquire knowledge. In pure fields such as philosophy or science, the pursuit of knowledge is ideally an open-minded affair engaged in for its own sake. In other activities such as applied research or industrial development, the inquiring mind seeks out new knowledge to support specific objectives.

The open-ended nature of research endeavors and their lack of obvious solutions and promising avenues usually qualify them as difficult problems. This chapter explores a number of domain-independent issues and techniques for pursuing research in various disciplines.

For the sake of concreteness, much of our discussion in this chapter focuses on the academic environment of graduate research. However, most of the topics and approaches pertain as well to research in other settings, whether a government laboratory or a corporate marketing office.

Agenda for Research

The newcomer to the research enterprise tends to have a simple view of what research entails. He initially believes that following a few suggestions from the advisor will lead to demonstrable results, a series of advances that can be measured by the hour or week. He has a tacit belief that progress will ensue automatically over time, much like attending a hamburger stand, raking autumn leaves, or solving homework problems. How could he believe otherwise? He has little or no prior experience with difficult problems of the magnitude that now face him.

Perhaps the most important thing he will learn in the first year or so is the environment of research. Addressing difficult problems requires a new mind-set, a willingness to explore new horizons, maintain an open mind, appreciate small insights, and even enjoy the steady stream of failures as well as successes.

Learning to conduct research systematically will be the most important aspect of his education in the first year. If the research effort spans a planning horizon of about two years, the indoctrination will occur in conjunction with an *orientation* phase during the first half-year, a period for defining the problem and gathering relevant information. If the orientation stage has been adequately addressed, the second half-year will be devoted to obtaining *preliminary results*, a set of tentative conclusions that may later be expanded, reinforced, or refuted.

The major results of the research are obtained in the third half-year. In this period of *substantive results*, the preliminary hunches and conclusions are augmented, refined, or replaced by more concrete results. The last phase, *wrap-up*, involves rounding out the collection of empirical data or cleaning up theoretical results. This is done in conjunction with the writing of the dissertation. New researchers invariably underestimate the amount of effort required for proper documentation, both in terms of describing the results in presentable prose and for ancillary activities such as the preparation of tables and figures.

Table 7.1 shows a representative timetable for a 2-year research effort such as a graduate program involving a thesis. The key difference for a longer program would be the elongation of the third "half-year," substantive results, into some indefinite multiple, whether two or five times as long.

In the early phases of the work, the researchers have an implicit belief that the stage of substantive results is the dominant component of the entire research effort, a phase stretching from the first to the last month of the timetable. They do not recognize its position as solely one stage in a larger scheme of things. In particular, the novice does not realize that there is something to learn about research strategy (Phase 0) or that the wrap-up stage (Phase 4) will consume so much time. It is better for the student to have a sense of these proportions at the beginning rather than the end of the research effort.

Selection of a Topic

Many a graduate student has spent months—and sometimes years—agonizing over the selection of a problem, then of the correct approach to tackling the subject. The

Table 7.1. Stages of work for a two-year research effort.

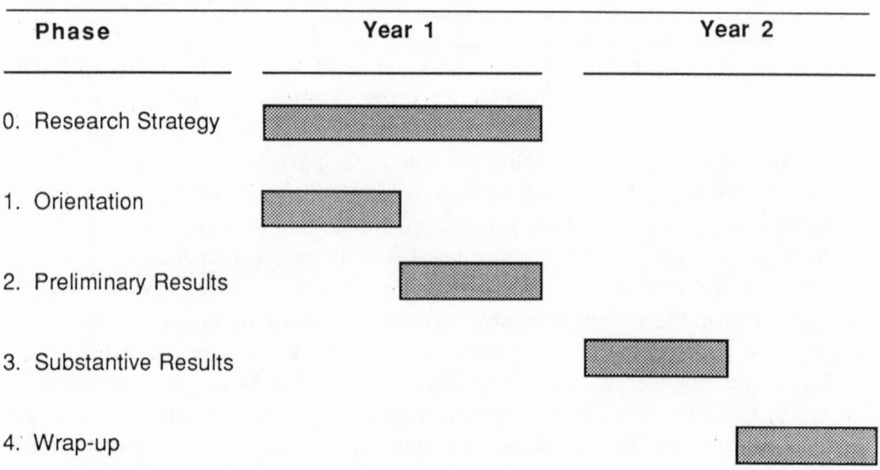

Phase	Year 1	Year 2
0. Research Strategy	�	
1. Orientation	�	
2. Preliminary Results	�	
3. Substantive Results		�
4. Wrap-up		�

world about us teems with challenging problems, and there is also an infinitude of easy problems. However, the intersection of those two sets—worthwhile problems that we can solve—often appears empty.

The appropriate topic to address will depend on the discipline, whether in astrophysics, economics, or chemical engineering. The choice of a worthwhile topic is a critical aspect of the research endeavor, as well as intelligent behavior in general. In the words of the mathematician Jacques Hadamard:

> How are we to select subjects of research? This delicate choice is one of the most important things in research; according to it we form, generally in a reliable manner, our judgment of the value of a scientist.
>
> Upon it we base even our judgment of research students. Students have often consulted me for subjects of research; when asked for such guidance, I have given it willingly, but I must confess that—provisionally, of course—I have been inclined to classify the man as second rate. In a different field, such was the opinion of our great Indianist Sylvian Levi, who told me that, on being asked such a question, he was tempted to reply: Now, my young friend, you have attended our courses for, say, three or four years and you have never perceived that there is something wanting further investigation?

This quote highlights two things: the importance of selecting the right question and the need for self-sufficiency. While I would agree whole-heartedly with the first theme, the second item deserves some qualification. Hadamard presents a strong view of the desirability for independence on the part of a graduate student. In reality, the beginning researcher should not hesitate to consult with her supervisor, since the role of the latter is to provide guidance. This is especially true in the early stages of the research effort. The student should meet with the advisor on a regular basis to assess her strengths and weaknesses, explore possible directions for research, discuss potential methodologies for tackling the candidate problems, and eventually settle on a tentative agenda for completing the work.

The discipline of inquiry will also constrain the acceptable methodologies for pursuing advanced research at the graduate level. A conventional dissertation in topology is likely to be all theory and no practice; one in urban planning may be mostly design and little analysis; and one in economics is as likely to focus on abstract theory as data analysis, computer simulation, or field studies.

Theory vs. Application

In many disciplines, a canonical approach to a good thesis is a judicious mix of theory and practice, of analysis and synthesis. This requirement, often implicit by precedent rather than explicit by fiat, may also take the form of evaluation and design, or of conjuncture and empiricism. This is as true of mechanical engineering as of materials science, and of linguistics as of computer science. The empirical component may take diverse forms depending on the domain, from the controlled experiment of the chemist to the participatory experience of the social psychologist and the remote observation of the astrophysicist.

The object of analysis, as the focus of the research, depends entirely on the field of investigation. It may be as impalpable as a neutrino or as concrete as concrete. It may be as small as a quark or as big as a galaxy, as self-contained as an axiom or as expansive as a society.

The object of synthesis may also take a variety of forms. It may be as tangible as a mobile robot or as abstract as a grammatical structure, as esoteric as a cosmological model or as prosaic as a computer program.

In the age of information, the computer has come to play a peculiar dual role in research, as both the product and the instrument. The hardware is the object of study for a computer engineer, and a tool of analysis for the statistician. The software is the object of synthesis for the knowledge engineer, and a tool of synthesis for the architect. With increasing frequency, the development of a good computer program suffices for the design component in many fields of graduate research. The software serves as constructive proof of the validity of the underlying approach.

In certain applications such as knowledge engineering, a single software package may serve as both the object of synthesis and of analysis. This happens, for example, when a newly constructed package is evaluated in terms of performance and efficiency, as a function of knowledge representation techniques.

When Is an Idea a Thesis?

A thesis as a hypothesis is an idea, but a single idea does not constitute a thesis in the sense of a dissertation. A beginning graduate student often regards a thesis as a single monolithic idea, and is thereby overwhelmed before he even starts. In reality, a dissertation is not a single great idea but a constellation of small ideas surrounding a common theme. In this sense it is no different from any other store of knowledge in science or engineering.

Suppose that these are the infant years of the twentieth century. A young physicist has been ruminating over the photoelectric effect: the ejection of electrons from atoms depends only on the frequency of the incident light, and not its amplitude. The prevailing model of light as a wave is inconsistent with the observed effect. Suddenly, an idea flashes into the scientist's mind—the quantization of light in the form of discrete packets rather than continuous waves. This alternative view would explain the paradox.

This insight would later lead to a Nobel Prize for Albert Einstein. But the idea, by itself, would have constituted a poor dissertation or article. More specifically, the central idea must be extended in the backward direction to determine its compatibility with previous theories and observations, and in the forward direction to assess its implications. If the idea were compatible with existing knowledge, then the relationships would have to be explained; if not, then the discrepancy would have to be justified, or the idea itself modified. Even seminal insights must be constructed on a bedrock of supporting ideas.

A Canonical Research Strategy

As alluded to above, the appropriate strategy for graduate research depends on the department and even the particular supervisor. For the sake of concreteness, a specific approach to the research task is outlined below. The novice researcher may find it useful as a reference guide, a point of departure for tailoring an agenda to her own discipline and research topic.

A canonical architecture for a good thesis or paper is as follows: Part I on theory and Part II on implications. This pattern is a stalwart—but not unique—structure for a good thesis.

Part I refers to the conceptual foundation, terminology, and general principles relating to an application domain, whether in economics, physiology, or sociology. The "implications" for Part II refer to theoretical results and/or practical applications—including case studies—of the general theory. Since the general framework and principles in Part I are inferred from some initial knowledge base, it is difficult or even impossible to write Part I before Part II. In practice, Part I is developed concurrently with Part II.

Theory and applications are developed synergistically in an evolutionary, iterative fashion. The researcher begins with some vague notions or flashing insights that are nurtured or eliminated through experiments or case studies. These results then serve as reference points for a new cycle of explorations. In fancier terminology, this is the stuff of the scientific method: observation (what you already know and learn afresh), induction or generalization (forming some vague hypotheses based on observations and/or case studies), and confirmation (validation of the hypotheses through further case studies).

The path to enlightenment is a long and winding one. In the rare moments of intellectual destitution and despair, the weary traveler may find consolation in the adage, "If you know what you're doing, it's not research."

- Pick some directions and destinations that seem interesting to you.

- Explore these directions. But you would not want to emulate the eager young man who jumped on his horse and rode off in all directions. Be modest: pick only two or four directions at a time, not twelve.

- Explore them in sufficient depth to pick out a few nuggets ("positive results"), or decide that there are none ("negative results").

- Keep in mind that a negative result is as effective as a positive result (see Chapter 6 and Appendix A).

- The thesis often discusses the paths that led to nowhere as well as others that went somewhere, in order to serve as a guide to future researchers.

- An active failure is actually a success, as discussed in Chapter 6 on the management of projects. The only thing to avoid is passive failure born of inaction.

As discussed previously, a thesis is a collection of small ideas focusing on a common theme, rather than a single monstrosity. One idea does not make a thesis.

- Plan on making some progress in the research each week, keeping in mind that an active failure is actually a success.

- Do not wait for the grand, perfect idea to streak into your consciousness; it will never come. Perfection is fiction in research as in business and everyday life.

- Explore your subject from different viewpoints, forging into neighboring disciplines that seem most relevant.

- Write down the positive and negative results at regular intervals. For theoretical results, this should be done at least once a week. An experimental setup or a design project can progress without weekly write-ups. But for theoretical or conceptual work, paper is the only product. Hence: no write-ups, no progress.

- Prepare memos of the explorations, both positive and negative, to serve as progress reports each week. These should be typed and not merely handwritten. For complex material such as theoretical discussions, the mind cannot grasp the entire argument when it is busy deciphering semi-legible text. Typed copy, on the other hand, will highlight the gaps and errors in argument.

- Writing the dissertation—as opposed to conducting the research—is a time-consuming, labor-intensive project. It is easy to underestimate the amount of work involved, and nearly everyone does.

- The progress reports will serve as a backbone for the final dissertation. It is amazing how much the mind can forget over the course of a year or two, a fact that will not hit home until the thesis-writing stage. Fortunately, much of the dissertation may be compiled verbatim from various sections of the progress reports.

Conducting research and writing about it are synergistic activities. For this reason, the prudent researcher will begin writing the first draft of her report or dissertation as soon as she has some coherent background information to describe, even before starting her first test, experiment, or field study.

The last phase of the researcher's task is to arrange her gems of discovery into a coherent piece of jewelry. To the experienced researcher, this is the easiest part of all. It represents the culmination of the research effort and is the most satisfying phase of the work.

When all is said and done, research can be an exhilarating experience. It may be so compelling that the researcher will cut short her vacation to get a head start on the next project! In the words of Albert Einstein:

> One of the strongest motives that lead men to art and science is escape from everyday life with its painful crudity and hopeless dreariness, from the fetters of one's own ever shifting desires.

In research the journey is as much the reward as the destination.

Summary

Research is often associated with university campuses, industrial research parks, and governmental laboratories. However, many of the issues and strategies for dealing with research relate to all walks of life, from gathering marketing data, preparing a speech, or culling notes for a class composition.

A case in point is the development of a new product. In product innovation, whether for a novel automotive engine or a convenient, multipurpose glue, the path to the goal is apparent only in retrospect. Even a working prototype is only an intermediate step, to be followed by the next challenging task of marketing the product effectively. A systematic approach to these problems requires a series of generalized observations and specific trials, a process that parallels the practice of scientific investigation.

- Research in execution often outlasts its planned time frame.
- The selection of a good research topic—challenging yet solvable—is half the battle.
- A thesis is not a single great idea, but a collection of several small ideas surrounding a common theme.
- A canonical structure for a good dissertation or paper comprises both theory and applications.
- "Negative" results are often as useful as positive outcomes.

8

Supervising the Researcher

Despite the pleasure that individuals obtain from their work, they are typically embarked on a solitary voyage, where the chances of failure are high. . . . It requires a strong constitution to go it alone in creative matters, and most innovative people at times experience a strong need for personal, communal, or religious support.   Howard Gardner

In the first sixteen years or more of our formal education, there is little to prepare us for the rigors of research or the demands of life in general. In lectures we are taught facts and techniques; in homework we develop skills by applying those techniques.

Even in project-based courses such as those sometimes found in engineering and business curricula, the experience is relatively structured. In general, the goals are precisely defined as are the alternative paths to the solution. Although more helpful than lectures, such project-based experiences still provide an inadequate preview of the rigors of earnest research.

There are courses in logic offered by the philosophy department, cognitive processes in psychology, and artificial intelligence in computer science. But they are not usually core requirements in the college curriculum. Further, even these courses generally deal with facts, figures, and straightforward deductive procedures.

These analytical and deductive methods are necessary but insufficient for solving difficult problems. The most challenging problems are, by definition, not straightforward. We are not taught in school how to grope intelligently, to stumble with style.

Our educational system, like society at large, discourages creative behavior which necessarily deviates from the norm. The forces of convergence, including the need for group identification and the fear of ostracism, are more numerous and powerful than those of divergence. Teachers, parents, and peers tend to encourage standardized rather than unexpected behaviors. The creative person must have a healthy dose of confidence and self-respect, since risk and creativity go hand in hand.

If we learn to think effectively and address difficult problems systematically, our skills spring from personal experience rather than formal education. For our educational system teaches advanced thinking skills in spotty fashion, at best. If we learn to think effectively, it is usually a by-product rather than a keystone of the course work.

Studies of 301 historical figures born since 1450 indicate the dubious impact of education on eminence. The sample included 109 leaders, ranging from the American general Philip Henry Sheridan as the most obscure to Napoleon Bonaparte as the most renowned; and 192 creators ranging from the English novelist Harriet Martineau to the French writer Voltaire. The results are depicted in Figure 8.1.

Perhaps the declining relationship between political leadership and education is not surprising. Intellectuals tend to have little flair for, or interest in, statecraft. A more curious result is the concave shape relating to writers, philosophers, scientists, and other creators. Eminence increases with education, yielding a peak at a college-level education just short of a bachelor's degree, then declines thereafter.

The educational process can and should be a vehicle for promoting creative behavior. The American psychologist Catharine Morris Cox offers the following conclusion based on her study of the 301 historical figures:

> Heredity sets limits, but within these limits the adequate training of the most gifted—and so also of their less distinguished fellows—may raise them to the designed stature of men unmarred by the defects of insufficient experience, and thus realize in each one the complete development of inborn worth.

According to the data, extensive formal education seems to be detrimental to the perceived eminence of creators in diverse fields. As with Albert Einstein during his childhood years, actual talent may be inversely related to academic performance.

> Time spent in pursuit of academic honors is time lost for the acquisition of information and expertise not directly related to schoolwork. It is time that also cannot be used for profound reflection. Many of the eminent are deeply involved in their own

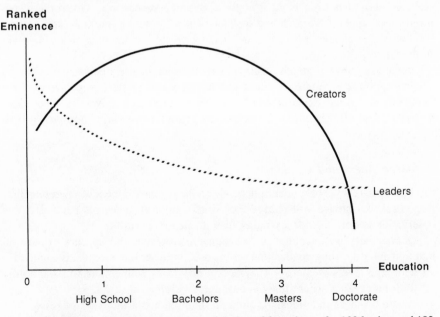

Figure 8.1. Relative (ranked) eminence as a function of formal age, for 109 leaders and 192 creators. [Adapted from Simonton (1984), p. 65.]

programs of self-education. . . . Research on the creative personality often points
to the importance of wide interests, a breadth of perspective, and a need for
novelty, diversity, and complexity.

Many other individuals, of course, regard academic competence as a necessary fact
of life and pursue it with vigor. But grades should never be allowed to displace a
stimulating program of self-education.

If a mission of formal education is to prepare students for a life of creative
contributions, then this goal has been poorly served in the past. Many have noted
that higher education forces students to know everything about nothing. Yet even
this view may be somewhat optimistic: "everything" seldom includes systematic
methods for tackling difficult problems, even in the respective specialty. Although
facts, figures, and domain-dependent rules—whether in music, art, or science—are
key ingredients of creativity, our formal education offers little in the ways of
synthesizing the pieces into worthwhile wholes. Our educational system should be
designed to nurture rather than squelch creativity. This may be done by encouraging
interdisciplinary pursuits, promoting independent research, and addressing difficult
problems that allow for multiple viewpoints and solutions. An effective vehicle for
implementing these goals lies in a research program of challenging proportions.

The task of research supervision remains a gray art. Although the high-level
topic of project management is a popular area of study in engineering schools, and
technological strategy a cornerstone of many business schools, the subject of the
first-line supervisor managing the researcher remains largely a mystery. "Ask a
dozen economists about economic policy and you'll get a dozen different answers"
goes the adage. If this is true for economics, the most "mature" of the social
sciences, it applies even more strongly to research supervision. Given the recognized importance of research and development to economic progress and societal
welfare, the subject of managing and leading creative individuals richly deserves
further study.

In the meantime, research supervision remains an intensely personal affair. This
chapter explores a number of ideas and methods that relate to my own research
environment, an interdepartmental laboratory at a research university. The reader
may find these ideas helpful as a point of departure for her own philosophy of first-
line management.

Resourcefulness and Guidance

The appropriate level of guidance depends on the resourcefulness of the researcher.
In general, the novice will require more intervention than the old hand. But this
observation is only a tendency rather than a categorical result.

Occasionally we encounter a newcomer to research who appears to have an
intuition about tackling challenging problems. These are often inquisitive individuals with high levels of energy. They accept new tasks with enthusiasm and boldly
set forth into realms where few have ventured before.

Where do these individuals acquire an intuition about the ways of research? I
suspect that their sense of ease springs from analogous experience in other domains.
The experience may be direct, as in the pursuit of challenging hobbies; or indirect,

as in reading biographical accounts of problem solving. Other sources of indirect experience might relate to a taste for mystery novels or science fiction, or perhaps even contemplation about the true character of scientific investigation and technological progress.

The individual with an immediate affinity for research is, unfortunately, rare. With few exceptions, the newcomer is also the novice. Consequently the beginner has to be led through a period of apprenticeship. As with most experiences, the first year of the research activity results in the greatest learning concerning the nature of research and strategies for handling unruly tasks. The subsequent years will tend to show decreasing rates of return; but often, enough will be learned during the first year to qualify the researcher as a full contributor to the project.

Structure

Given the general lack of experience among novice researchers in tackling difficult problems in a systematic fashion, the task of supervising research often requires as much creative effort as the underlying problem itself. What is needed in supervision is a careful balance of guidance and distance, of control and delegation. On one hand, it is possible to be overbearing:

> Some companies give employees the freedom to use a percentage of their time on projects of their own choosing, and set aside funds to explore new ideas when they occur. Others control resources so tightly that nothing is available for the new and unexpected. The result is nothing new.

On the other hand, it is possible to promote an atmosphere so relaxed as to eliminate all incentive and motivation. The absence of incentive may result as much from lack of understanding of the ultimate vision or problem, as from the lack of perceived reward.

Structure can often be a vehicle for creative expression rather than a straightjacket. The sonnet specifies tight constraints on the length and pattern of each line, but has engendered some of the most beautiful poetry ever written. Most programming languages offer only a few dozen key words and prescribe rigid constraints for forming acceptable statements; but they can express any kind of rational procedure or computation that we may wish to encode.

In general, structure and guidance can serve as enablers rather than limiters of creativity. A study of innovation in American companies by the management specialist Rosabeth Moss Kanter yields the following conclusion.

> True "freedom" is not the absence of structure—letting the employees go off and do whatever they want—but rather a clear structure which enables people to work within established boundaries in an autonomous and creative way. It is important to establish for people, from the beginning, the ground rules and boundary conditions under which they are working. . . . Without structure, groups often flounder unproductively, and the members then conclude they are merely wasting their time.

For these reasons, the benefits of a laissez-faire policy must be balanced by the need for some structure in directing creative energy. The need for structure will depend on the individual personalities as well as experience. In general, the more experienced the individual, the less structure required.

The research endeavor appears long and arduous to the graduate student. The supervisor should obviously help the student along this journey through both intellectual and emotional support. He should provide guidance and offer rapid feedback on the efforts made by the student. He should offer encouragement and praise whenever appropriate.

On the other hand, it is possible to proffer too much recognition. Some students tend to relax at the first sign of positive reinforcement from the supervisor, even when the praise was intended largely to encourage the student whose performance still falls short.

In general, external influences—whether praise or criticism—may be detrimental to the quality of creative results. Studies of scientists and innovators indicate two necessary conditions for highly creative work: (a) the individuals should regard the work as intrinsically interesting or personally satisfying, and (b) they should not be hampered by extrinsic factors.

While performance on easy tasks may be enhanced by external rewards, open-ended tasks require intrinsic motivation. This observation has been depicted graphically in terms of working through a maze by the American psychologists Beth Hennesey and Theresa Amabile. If the primary motivation lies in some reward conferred at the completion of the maze, then the compelling strategy is to rely on previous experience and negotiate the territory in minimal time, at minimal risk. An intrinsically motivated person, on the other hand, actually enjoys being in the maze. As a result, he will explore it more fully by following unknown routes, and will ultimately emerge through a novel, and perhaps more creative, pathway.

The task of the supervisor, then, is to motivate his charge by sharing his vision and imparting a sense of the joys of discovery. To do this, the supervisor must shield the supervisee from administrivial distractions, handle external relations, and ensure that promising developments or successful results are adequately communicated to the external world. This insulation must be partial, since complete isolation would correspond to lack of feedback and guidance.

Supervision of research may be one of the most challenging forms of leadership, as the supervisor must inspire action at a distance. In guiding the researcher, the supervisor must provide motivation and feedback, but only so much. And this is no simple task. Despite its challenges and pitfalls, however, the role of the research supervisor offers its unique rewards.

Selecting a Problem

The world around us offers a cornucopia of problems, some of which are tractable and others not. Perhaps the most important role of the supervisor is to help the researcher find a worthwhile and manageable problem by matching the needs of the proposed project with the researcher's personal skills and interests.

If the mission of a research group is to advance the technology of factory automation, then the subgoals range from intelligent robots to communication protocols, from machine tools to transport devices, from process scheduling to production management, and from process physics to system integration. These are some widely recognized areas of research in factory automation, but perhaps the supervisor recognizes the need for addressing an entirely new set of topics.

For any topic, there is usually a variety of tools and methodologies that may be applied. Consider the goal of developing intelligent robots. Some referent disciplines for this task include information theory for investigating signal processing efficiency, dynamics for an evaluation of forces on joints, control theory for stability considerations, materials science for improved weight and strength, statistics for reliability analysis, artificial intelligence for object recognition, computability theory for software efficiency, and automata theory for theoretical limitations to automation. Here, too, the supervisor may see the need for an entirely new perspective, such as the promise of a theory of parallel processors to model the numerous activities, both mechanical and computational, that occur concurrently in any robot possessing a modicum of intelligence.

In the early phase of a novice researcher's tenure, the supervisor can do little but provide exposure to the topics. The new investigator should explore the topics of personal interest: he may be keen on production scheduling and machine tools but bored by all other topics. Often the researcher is not entirely certain of his interests beforehand; five weeks into the exploratory phase of production scheduling, he may decide that factory layout is his true love.

Once a tentative topic is selected, the team must formulate a general strategy for dealing with it. The strategy will include the methodology, whether in terms of laboratory tests, hardware prototypes, software packages, or theoretical models. Often a combination may be employed; an example in the engineering realm would be the development of a theoretical model, followed by computer simulation and construction of a prototype. In the social sciences, a theoretical model of a problem domain might be followed by field studies and statistical analyses.

An important component of the research strategy is the identification of appropriate tools and techniques. As discussed above, often more than one technique is relevant. Here, too, the supervisor must help to match the tools to the supervisee's skills and interests, based in part on the researcher's previous experience and future goals. The researcher may be a wizard in computer programming but not mechanical design; or in statistics but not differential equations.

If his skills are relevant to the selected topic, then this may mark the beginning of a happy ending. But what happens when a researcher has a burning desire to study quality control in production lines but has little aptitude for statistics? This is where the supervisor has to weigh the needs of the project against the the training or education of the researcher. If the researcher has an interest in other topics which seem almost as appealing, then the solution is simple: select another topic more in tune with the person's aptitudes. On the other hand, if the researcher has convinced himself that he wants to spend the rest of his working life in quality control, then the case is more involved.

The supervisor can opt for one of the following: agree to let the researcher pursue his inclinations and ignore the lack of progress on the project, or work with the supervisee to develop a new set of techniques that apply to the topic. The road traveled will obviously depend on the strengths of the passions involved.

In the exploratory phase of problem selection, the supervisor should provide as much leeway as possible and help the researcher find a topic that suits his tastes. Otherwise, pursuing a topic that is of little interest to him will only result in his perpetual daydreaming about the romantic project that might have been.

The identification of a problem and a corresponding research strategy sets the stage for the active phase of the research. In this phase, the supervisor can afford to be lax only with highly motivated supervisees. For the others, he has to be more assertive. To adopt a laissez-faire policy may be emotionally appealing, but the novice researcher left to his own devices will only spin his wheels and become discouraged. The uninvolved manager whose sole priority is to play Mr. Nice or Ms. Glucose, offering vacuous words of encouragement devoid of intellectual substance, is of little help to the researcher. In graduate school, students often waste many months and even years roaming aimlessly across the intellectual landscape, exploring unrelated topics here, there, and everywhere.

Occasionally, the itinerant explorations may yield ideas that eventually coalesce into a superlative dissertation. Some highly motivated students would consider the wanderlust an integral and worthwhile part of the research experience. But others would surely give up some of the soul-searching to have more active guidance from their supervisors.

Obviously, "active" guidance does not mean "unerring" counsel. The path to a first-rate dissertation is rocky and risky. The supervisor, too, will fall prey to the mistakes that are the spice of creative problem solving. The message is not that the supervisor should strive to be superhuman, but that he be more than a passive conduit for conveying administrative requirements for a graduate program. With many students, the supervisor should actively guide the work, leading his student over hurdles and around obstacles.

As discussed previously, the results of research are almost always presented in a coherent, systematic format. The novice researcher is thereby misled into believing that research itself proceeds in orderly fashion. In reality, any research of respectable difficulty proceeds forward as well as sideways and even backward.

The beginning researcher notes the discrepancy between the coherence of reported work and his own progress, or lack of it. He then comes to feel that his project completion date is a moving target that is always N months away, for some N equal to 2 or 20, depending on the project.

The novice researcher often has only a hazy perception of the objectives and is even more unclear about the methodologies. He is stultified by the magnitude of the problem and refuses to take small steps, despite suggestions and entreaties from his supervisor. To him, the suggestions represent short-term ramblings in all four directions that appear unlikely to amount to anything. The research group is surrounded by a mountain range of goals that loom in the misty distance, but no clear path exists to even one of the peaks.

Failure can often be more rewarding than success itself. Consider the following vignette.

> I could almost feel the flush of success. One more day, and we would reach Diamond's Peak. According to legend, diamonds were as plentiful as ice at the summit of this, the highest mountain in the world.

Faith and I would gain not only wealth, but fame, for no one had ever returned from an odyssey to the peak of peaks. Our mortal enemies in this endeavor were wind and snow. The blistering wind, raging at speeds approaching two hundred kilometers per hour, screamed continuously in our ears and blocked our vision with sheets of driven snow.

In the gathering dusk, we secured the provisions and tethered our sleeping pouches amongst three large boulders. As I lay in my pouch pondering over the following day's journey, I began to feel an alien presence, a disturbance in the ether which grew with each passing moment. This eerie sensation was presently joined by the whistling of an enchanting melody.

I turned my head to face Faith. She formed silent words with her lips.

"What was that?" she asked.

"Who goes there?" I cried out.

But my words were garbled by the din of the maelstrom. As if in response to my cry, however, a furry creature, perhaps three meters tall, leaped over the smallest boulder and bounded over our pouches toward the supplies. In one smooth motion it ripped away the cords and garnered both the equipment and provisions. In the next instant it was gone.

Could it be some trick of the mind in a moment of excitement and stress? I turned once more to Faith.

"Did you dream what I dreamt?" I feared to hear her response.

"Are hallucinations contagious?" she replied. Then her usual good cheer gave way to a wistful look. "But I do believe our supplies have walked away."

"With an apparition?"

"Must be the Snowman."

In hindsight, it seems obvious that we should have prepared for an encounter with the Abominable Snowman. Yet . . . yet there had been no reports of the icy menace in this part of the world.

Without food or equipment, Faith and I could not continue with the final ascent. There was no path but to return to the base camp, and that as rapidly as possible.

Success, which seemed so close at hand, was dashed by circumstance on this ill-fated day. But no matter—next time we would be better prepared.

This brief yarn offers several morals. First, no amount of planning will fully prepare the traveler for a journey into uncharted territory or an encounter with a difficult problem. After a reasonable amount of deliberation, the traveler must forge ahead and actually get his boots wet.

Second, the sooner one tries and fails, the sooner one can succeed. Of course, the repeatable nature of a journey depends in part on the circumstances specific to the task. Nature lets us fall off a cliff only once, but is more generous about our falling into lakes or suffering intellectual blows.

Third, the failure, at the very least, adds spice to the original mission. And at best, it may suggest new directions of inquiry. How did Mr. Snowman get to Diamond's Peak? How does he survive? What does he do for entertainment? Where are his parents? Does he know where his children are? The new line of inquiry may take the voyager on a journey even more engaging than the original adventure.

The unknown "X" factor that turns up uninvited in a problem-solving effort is a regular participant in scientific discovery and technical invention. The best ideas often stem from explorations that go astray.

The discovery of penicillin is a celebrated example of fruitful diversion. The British bacteriologist Alexander Fleming was conducting experiments on a virulent strain of bacteria, the *Staphyloccus aureus* commonly found in boils and other skin infections. One early September morning, Fleming noticed that the staphylococcal colonies were absent from the region surrounding a large mass of mold that had accidentally settled on one of the petri dishes.

The scientist of average caliber would have considered this a ruined experiment, another small sacrifice at the altar of experience. Fleming, however, recognized significance in the failure. But when he showed this plate to colleagues in his laboratory, he was flabbergasted by their lack of interest.

Despite the reaction of his colleagues, Fleming initiated a series of experiments that continued on and off for over a decade. The new agent—penicillin—came from a variant strain of *Penicillium notatum*, a variant so rare among thousands of existing molds that Fleming's accidental discovery would have been "most unlikely" to be repeated. Penicillin became widely available in commercial quantities following the Second World War. For his work, Fleming was knighted in 1944 and shared the Nobel Prize for medicine the following year.

In the realm of graduate research, some students accomplish more on a course project requiring 5 hours of work per week than on thesis research to which they devote an entire semester. This paradox is related to motivation. For a young researcher, the problems that loom ahead are bigger than any homework problem— bigger, in fact, by orders of magnitude. Faced with this monstrosity of alien proportions, he feels small and inadequate. Typically, however, the researcher is a bright individual who previously excelled in academics; otherwise he or she would not have joined a research group.

What is the supervisor to do? The solution to the dilemma is to (1) partition the problem into a series of digestible chunks, each requiring no more than a few weeks of work, and (2) provide a series of concrete successes. Fortunately, these goals are positively correlated.

To bend an adage, "Nothing succeeds like success or failure." The investigator and supervisor must document their exploration along the way, so they can learn properly from mistakes and build on the successes. The documentation will serve as project milestones, and as laurels on which to rest. When enough laurels have accumulated, they constitute the fibers of the ultimate project documentation, such as a final report or a dissertation.

These progress reports are not only intended to fulfill the project objectives, but also the researcher's educational interests. Hence an important factor in their evaluation will be the amount of *demonstrated* effort, as opposed to clearly identifiable results.

A supervisor should note the emphasis on demonstrated effort that is reflected in the report. Examples of such effort are embodied in written responses such as the following:

- "I have read 4 books on information theory and here's how they seem to relate to our problem. . . ."

- "Attached is a set of tables that identify the salient attributes of the domain."

- "Figure 1 presents a tentative framework showing the relationships among the key concepts."
- "For reasons discussed previously, mathematical logic seemed to be a promising tool to use for formalizing our model. But it is actually not relevant for the following reasons. . . ."
- "The limitations of automata theory in describing only computational features, implies that we must extend this model to account for physical transformations by the following scheme. . . ."
- "In retrospect, our failure was inevitable. Under the prevailing conditions, the goal was impossible to attain for the following reasons. . . ."
- "A postmortem analysis of the theoretical assumptions and statistical variations in experimental apparatus indicates that the probability of success was 7.6 percent."
- "Our result lies within 5 percent of the theoretical maximum. Increasing the development budget and operating expense by a factor of 10 would raise performance by only an additional 3 percent. This task we leave to future investigators."
- "A refined model of the key elements and their relationships is as follows . . . ."
- "A preliminary architecture for the software is given in the attached figure . . . ."
- "The key aspects of the knowledge representation, plus the relationships among the critical modules, can be implemented in software as shown in the 2–page program in the appendix."
- "Since the program takes too long to produce a solution, I suggest the use of the following domain-specific control procedures. . . ."

It may be helpful for a graduate student, for example, to think of his thesis research as a series of semester-long "independent research" projects. The results of the research should be written up in a series of interim reports culminating in a Progress Report for the semester; a tentative outline is given in Table 8.1. These reports can be evaluated by the supervisor to provide additional feedback on the researcher's progress.

Often the novice researcher has a "creative" view of what constitutes effort. He can generate any number of rationalizations to explain why he is unable to make any progress. Here are some typical comments from beginning researchers, and the reality they attempt to mask:

- "I have read all 13 books on the subject available in the library. I 'understand' them, but I don't see how they relate or don't relate to our work." (The researcher does not understand the material in sufficient depth, if he cannot even explain why it is *not* relevant.)
- "That's too easy. I don't want to waste my time writing such silly little programs. I could tackle real programs if you could just show me how to

Table 8.1. Illustrative outline for a progress report relating to conceptual or theoretical efforts

---

1. **Summary**
2. **Objective**
   • What is the purpose of the work?
3. **Background**
   • Review of literature
     • What is the nature of similar work in the past?
   • Relevance
     • In what ways and to what extent is existing knowledge relevant to the objective?
   • Irrelevance
     • In what way and to what degree is previous work not relevant?
4. **Approach**
     • What is the conceptual framework for the problem domain?
     • What methodology or techniques were used?
     • What activities, operations, or experiments were conducted?
5. **Results**
   • Positive
     • What are some tentative or definitive results?
     • How can the current work be adapted, extended or otherwise modified to better meet the objective?
   • Negative
     • What is the mode and extent of each failure?
6. **Strategy**
     • Limitations of Current Approach
       • What are the drawbacks of the current approach?
     • Future work
       • What more needs to be done?
7. **Attachments**
     • References
     • Appendices
       • Technical details
       • Ancillary information
       • Computer programs
     • Figures
     • Other

---

handle some pesky details." (The researcher cannot progress in his work because his knowledge of basic materials is skimpy and tentative.)

• "I didn't look too carefully into that topic because it might not be useful." (How does he know that beforehand?)

• "I didn't follow your suggestion because I can't see how it relates to our project." (The novice cannot see the forest for the trees; if he did, he would not be a novice. Moreover, he will not shed his novice-hood in a hurry if he refuses to participate in the game of research.)

• "How can I get anywhere if you don't give me enough detailed suggestions?" (He has followed your initial suggestions only superficially, and is now inadequately prepared to implement your subsequent ideas. He wants you to teach him how to play winning polo, when he can hardly stay on the horse.)

- "Why should I spend more time on this approach when we know it won't work?" (The fact that an approach failed is not particularly illuminating in itself. But the manner and extent of the failure provides the supervisor with two important insights: (a) better understanding of the territory under exploration, and (b) improved perception of the skills, talents, and interests of the supervisee. Each of these is key to determining the subsequent tactic or strategy.)

The hallmark of the experienced researcher is the constructive use of failure. Almost by definition, failure is the chief characteristic of addressing difficult problems. The fact that a tactic failed does not in itself provide much insight. However, an examination of the ways in which the approach failed, and the extent of each shortcoming, is the foundation for success. These discrepancies may be analyzed to determine how the previous approaches may be refined, extended, or replaced to advance the work to a new plateau.

Assertiveness

How does one supervise a Nobel Prize-winning scientist? I imagine one attends to his needs then leaves him alone. Supervising a neophyte researcher, on the other hand, is a different matter. A measure of assertiveness seems to be called for.

A supervisor may be inclined to foster a spirit of camaraderie in his team, a sense of fellowship in the noble quest to push back the frontiers of ignorance. But reality has a tendency to intrude where it is least welcome. However worthy the atmosphere of liberty, equality, and fraternity, a policy of democracy sometimes fosters anarchy.

In a setting of complete equality, the novice researcher comes to feel that his hunches on the subject, uninformed as they are, are as trustworthy as his supervisor's—even more so, for they are his very own. Further, you as the loyal supervisor will surely fight to your death for his right to his opinions. And his hunches tell him that you don't know which way is north. So he is reluctant to pursue your suggestions, even when he has no alternative to offer. Meanwhile the suspense continues: time marches on and the research stands still at a furious pace.

The new researcher has little or no perception of what the research enterprise involves. How is he to respond to your queries along the following lines:

- "Would you care to look into the historical background of this problem?"

- "Could you find out more about information theory so we can figure out how it might relate to automated manufacturing systems?"

- "Will you write a 2-page program that incorporates the top-level modules of our software architecture, ignoring the details for now?"

In the darkness, the uninitiated can hardly perceive the tunnel that you want to explore, let alone any signs of light at the end of the trek. So he usually responds in one of the following ways:

- "No. I don't fully understand what the point is. Why bother to do this?" So you explain your reasons why, the hunches and the tinglings in your stomach.

Of course you cannot give a full, watertight argument, for otherwise there would be no need for exploratory research. Unfortunately, the novice finds this state of affairs entirely unsatisfactory, perhaps irrational. He might even wonder about your mental equilibrium. But after an hour of earnest convincing on your part, he mumbles, "Well . . . all right" and saunters off. But he returns the next week and you realize that he has done little in the interim but brood.

- "Yes . . . if you say so." He is more polite, but also unconvinced. The same scenario applies as above, except that you are spared your one-hour pep talk.

- "Yes. I'll give it a try." He is unconvinced, as usual. But he has nothing to lose, so he makes an effort. Since he cannot see the wherefores, however, the how-to of the whatnot is only half-hearted, and the whereto is short of fulfilling.

The supervision of creative talent requires respect for the supervisee and his feelings. This means maintaining a comfortable atmosphere conducive to the free expression of feelings, as well as tolerance for periods of silence during which the individual may be thinking.

On the other hand, it is possible to confuse abundance of respect with lack of direction. In my early years of supervising graduate students, I tried to foster a sense of equality in developing research strategy, a spirit of joint partnership in the great search for enlightenment. I have come to learn that this is good for morale in the short run, but not the long. When I assumed the role of partner rather than a guide or tutor, my students were haunted by the fear that all the tunnels I suggested might lead only to dead ends, not to freedom. If we were equals, then the student's opinion would be as valid as mine, and his overriding conviction was that each course of action would be futile and therefore not even worth trying. So the project would stand still, as would the student's dissertation. In this setting, the dawn of hope turns into the dusk of dismay without seeing the light of day.

The better approach is to be more assertive in the early phases of the research effort. More fruitful are statements like the following:

- "You should look into the historical background to this problem . . . ."

- "Let's find out more about information theory. . . ."

- "Please write a 2-page program. . . ."

rather than "Would you care to . . . ," "Do you suppose you might . . . ," "How about the possibility of . . . ," or any number of tentative statements.

Surely every research supervisor has received comments from former supervisees along these lines: "At first I didn't know why you wanted me to try those crazy things, but it makes some sense now." Or, in a similar vein, "In the first year I did what you said only because you insisted; in retrospect, there's a trace of method in your madness."

Only by participating in the game of "two-steps-forward, one-step-back" can the neophyte learn to appreciate the significance of productive failure. As the months roll by, the novice gradually turns into an expert in his own domain. With increasing practice and savvy, the researcher can then play a more active role in the formulation of the research strategy as well as its implementation.

Motivation

Psychological studies in laboratory settings indicate that tangible rewards may be detrimental to performance. If a subject finds intrinsic interest in the task *and* the problem is difficult, then external incentives will often decrease her performance. On the other hand, if the task is aversive to the subject or the solution is straightforward, then rewards tend to enhance performance. To borrow an observation, "Interest in challenge is like a delicate exotic flower; touch it and the bloom is gone." An implication for the supervisor is the following heuristic:

> **Laissez-faire Principle.** Do not disturb a genius at work, whether through positive or negative reinforcement.

If a supervisor is privileged to work with someone who is spontaneously creative, his best policy is to shield the creator from unproductive inputs, whether from the world outside, or within the organization itself. This protective role was the key contribution of Tom West in the development of a new-generation computer at Data General.

As discussed in Chapter 5 on methodologies, a researcher needs to acquire knowledge relevant to his work. Hence the supervisor must take care that in his enthusiasm for protecting his team from distractions, he still leaves open channels and sources that may inspire the researcher.

On the other hand, it is not every day that we come to supervise creative geniuses who can work productively largely on their own. In that case we need to intervene more directly, by participating actively in the work itself as well as providing feedback on the supervisee's progress.

Summary

Our formal education arms us with numerous tools for addressing well-defined problems with straightforward methods of resolution. Unfortunately, most of life's problems are neither clearly specified nor tightly constrained. For the most part, these problems we learn to address through trial and error plus an occasional insight from the experience of others. Since the research enterprise is more akin to the mushy problems of life than to class assignments, the novice researcher must be initiated into its dark arts over the course of his apprenticeship.

- The appropriate level of guidance varies inversely with the resourcefulness of the supervisee.

- Structure can serve as a vehicle for, rather than an impediment to, creative expression.

- A key task of a supervisor is to motivate the supervisee.

- An effective strategy for overcoming an obstacle is to partition the problem into a series of digestible chunks, then provide for a series of intermediate successes.

- The results of any research completed should be written up in a series of interim reports.

- A supervisor must determine the appropriate levels of assertiveness and motivation for guiding each supervisee.

9

Conclusion

The world is but canvas to our imagination.   Henry David Thoreau

Previous approaches to creativity have often focused on the person or problem domain, as well as the task itself. In this book, we have focused on the task: a difficult problem is one that has no ready solution or even the means to a solution. Some consequences of this perspective are as follows:

- Creativity is a matter of degree. The operant question is not "Is this result creative?" but rather *"How* creative?"

- Creativity is a domain-independent concept. An accountant may be creative, as may a shopkeeper or a musician.

- All of us face difficult problems from time to time. We may be creative at one point, and uncreative at another.

- Creativity involves *purposive novelty*. Originality or diversity is a necessary component of creativity, but diversity in itself is not a sufficient factor if it does not resolve the referent problem.

- As encapsulated in the Multidistance Principle, the solution must incorporate components exhibiting some properties that are distant, and others that are close.

- If creativity is a form of higher-order problem solving, itself a cornerstone of general intelligence, then there exist rational approaches to enhancing creative results.

- An effective procedure for dealing with difficult problems lies in the Method of Directed Refinement.

- Active failure is the highway to success.

- Productivity in project managment involves the pursuit of a select number of parallel activities: too few, and efficiency suffers through slack time; too many, and overhead paralyzes productivity.

- Our social institutions, including the educational system, encourage conformity—a homogeneity often leading to mediocrity rather than harmony. For each problem of consequence, we should rather seed myriad ideas and cultivate multiple solutions.

- Supervision of creative individuals is a delicate affair involving both intervention and insulation. It calls for inspiring action at a distance, without undermining interest nor tainting intrinsic drive.

In this book, we have partitioned the components of creativity into five factors: purpose, diversity, relationships, imagery, and externalization. The purpose of the creative effort defines the problem to be resolved. Since the solution is not obvious in a difficult problem, the result must contain some element of novelty or diversity. The components of the creative result must display some distant attributes, but must also exhibit certain relationships that bear on the task at hand. Imagery in its various guises, whether of sight, sound, touch, or taste, is a key factor in thinking about potential solutions. The importance of each type of imagery will depend on the person and the task, varying from the acoustics engineer concerned with automotive vibrations to the master chef concocting her next delectable.

Finally, the solution to a difficult problem must be expressed or implemented in some form. The need for externalization is both an end and a means: often the attempt to express thoughts in words, pictures, or models will accelerate the resolution of the problem. This may be due to several factors: the use of the extrinsic medium as an extension of working memory, the ability to step back and assume an objective stance, and the spotlighting of obvious limitations when the tentative solution is cast in clay.

A Future of Creativity

Too often we travel down well-worn paths, choosing the security of familiarity over the risks of novelty. We select problems—whether in the arts, sciences, or personal affairs—because they are readily solvable rather than significant. I am reminded of the story of the Martian who encounters a fellow star-trekker on the deserts of Mercury:

> "Good heavens, Marty! What are you doing in this oven?" asks the first creature.
> "I'm looking for my contact lens," replies Marty.
> "You're sure you lost it here?"
> "No—on the other side of the planet."
> "Then why crawl around in this life-forsaken place?"
> "The light's better here!"

This small book explored the premise that difficult problems are not always impossible, and methods of resolution not magical. We contemplated the joys of tackling difficult problems, and the rewards of their resolution through strategies such as directed exploration.

Philosophers, critics, and visionaries have at various times expounded on the need for active progression at all levels of society. Their arguments include the expansion of human spirit through challenge, the withering of individual character through complacence, and the death of collective culture with lost frontiers. But the absence of frontiers is only an attitude, the product of myopic vision or bankrupt policy, for uncharted worlds abound within ourselves and without.

Enigmatic problems stunt our spirit in the short run, but nurture it over the long. To be is not always to do, but to grow is to become. We could do worse than to face the future with optimism, to seek out problems of consequence, and to grow with the adventure.

Appendix A

Creativity in Science and Technology

The advance of science is not comparable to the changes of a city, where old edifices are pitilessly torn down to give place to new, but to the continuous evolution of zoologic types which develop ceaselessly and end by becoming unrecognizable to the common sight, but where an expert eye finds always traces of the prior work of the centuries past. One must not think then that the old-fashioned theories have been sterile and vain.

Jules Henri Poincaré

The scientific enterprise is one of generating new knowledge for its own sake. The technological process, in contrast, refers to the application of knowledge to satisfy human needs other than curiosity. The spectrum of technological activities, from science to engineering and marketing, is depicted in Figure A.1.

Types of Discovery

If the fields of science and technology are to progress in orderly fashion, it is important to develop a systematic theory for the nature of results attained in these domains. To this end, a framework is presented for the types of discoveries as well as their methods of derivation.

The process of scientific discovery is depicted in Figure A.2. An investigator, working individually or in concert with colleagues, envisions a result based on her knowledge of the universe. This knowledge may result from direct personal observation or indirect knowledge through the work of others.

The result is generated through some methodology. The specific method may rely on some cognitive mechanism we do not yet comprehend, such as the realization that sunrise and sunset are due to the rotation of the Earth, rather than the motion of the sun. We call these yet-unknown mechanisms "intuition" or "inspiration."

On the other hand, certain methods are more straightforward, as exemplified by the technique of proof by contradiction or refutation. To illustrate, we may show that a specific statement in predicate logic must be derivable from an initial set of hypotheses, by showing that the negation of the statement would imply some inconsistency in the overall set of statements. In fact, this refutation procedure is routinely used as the basis for the programming language of Prolog.

The results of scientific research may be classified into the following four categories: alignment, possibility, impossibility, and trade-off. These groups of results may be obtained by methods of construction or contradiction.

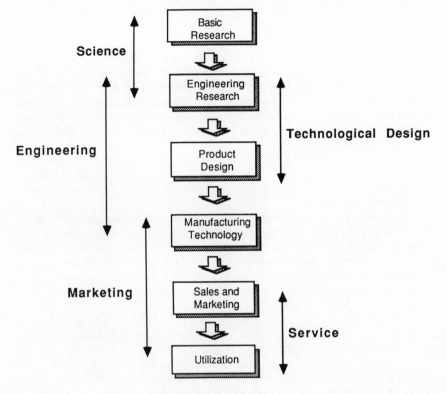

Figure A.1. The phases of technological development.

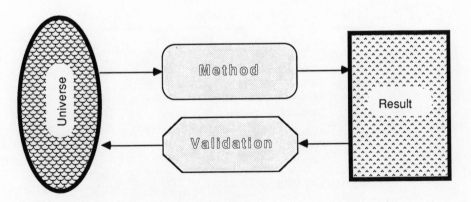

Figure A.2. The process of scientific discovery.

Table A.1 shows examples of results by category and method of proof. These classifications are discussed further below.

Alignment

Alignment refers to the fit between our models and the world around us, or among the models themselves. In attaining such harmony, our perception of the universe takes a simpler form. The issue of alignment may be further classified into two types: paradigm and unification.

A *paradigm* is defined by our perspective of the universe. Hence a paradigmatic result refers to a shift in our views. For example, the replacement of the geocentric view of the solar system by the heliocentric paradigm represented a major advance, and was instrumental for further advances in astronomy.

On the other hand, *unification* refers to the alignment among our models or views of the world. A unifying structure provides a general framework in which to organize results that previously seemed unrelated. The structure may take the form of a framework, model, theory, or some combination.

An example of a unifying structure is the development of the periodic table, and the classification of elements into related groups based on their electron configurations. Another example is found in the theory of electromagnetism, which unifies the seemingly unrelated phenomena of electricity and magnetism. This integrative model also accounts for many types of radiation, from gamma rays at one end of the spectrum to radio waves at the other. We now recognize ultraviolet emissions, as well as light and heat, as variations on the single theme of electromagnetic radiation.

Another unifying structure is found in the laws of thermodynamics, relating to the conservation of energy and the tendency of systems toward increasing disorder. These two laws encapsulate observations that arise in all realms of natural science and engineering.

A subcategory of unification relates to laws of invariance. *Invariance* refers to the constancy among objects that appear to be different at first sight. Such principles assert, for example, the immutable nature of certain objects despite transformations.

The conservation laws of physics typify the category of invariance principles. For example, the principle of matter-energy conservation states that matter and energy may change from one form into another, but the total energy of a system remains the same.

Table A.1. Examples of results by category and method*

| | Method | |
Category	Construction	Contradiction
Alignment	Heliocentric model Electromagneticism Central Limit Theorem	(Unknown)
Possibility	Intelligent machines Organic out of inorganic materials	(Not applicable)
Impossibility	(Not applicable)	Halting Theorem Perpetual Motion Machine
Trade-offs	Algorithms in Complexity Theory	(Unknown)

*Science tends to focus on alignment and impossibility, while engineering addresses possibility and trade-offs.

The single most important result in the field of statistics is the Central Limit Theorem. This theorem states that a particular probability law called the Gaussian distribution is pervasive. The Gaussian function, familiar to many people as the "bell-shaped" curve, serves as a good model in many practical and theoretical applications. A common example is intelligence scores, whose bell curve peaks at 100 and trails off toward either end to reflect the fact that decreasing numbers of individuals possess either very high or very low scores.

Engineers often rely on *dimensional analysis* to show the validity of their reasoning. This approach is based on the idea that the physical properties of systems may depend only on the combinations of certain characteristics, rather than their individual values. The field of fluid mechanics, for example, uses the Reynolds number:

$$R = \rho v d \;/\; \mu$$

where ρ is the density of a fluid having viscosity μ and flowing with velocity v in a. conduit diameter d. The units of these parameters are such that R is a dimensionless number. The fluid travels in orderly, laminar flow for low values of R, and becomes turbulent for high values.

Crude oil flowing in a transcontinental pipeline has a different character from water running through a garden hose. Their densities differ as well as their viscosity or internal resistance to flow; the diameters of the conduits will be dissimilar, and the speeds of flow may also vary. In other words, the values of the density ρ, the velocity v, the diameter d, and the viscosity μ are distinct for each fluid. But as long as the compound factor $\rho v d / \mu$ is much less than the threshold of about 1000, the flow will be laminar. On the other hand, if the number is much higher than this threshold, the flow will be turbulent. The identity of the fluid itself is of little consequence in this determination. Only the composite parameter in the form of the Reynolds number is a reliable indicator of turbulence, short of actually observing the fluid under the stated conditions of flow.

The field of automata theory uses a model of computation called the Turing Machine. This model depicts computational procedures as a set of simple operations. Much of the work in automata theory deals with the invariance of computational power among different versions of the Turing Machine. One such result is the equivalence of all existing computers to the Turing Machine—and therefore to each other—in the range of problems they can resolve.

Einstein's Theory of Relativity may also be regarded as an invariance result. In particular, the laws of physics are unchanged by the choice of a particular frame of reference.

Possibility

Much of the work in the sciences and in engineering deals with showing what is possible. One of the most convincing ways is proof by *construction*. The most cogent means of showing that humans can attain powered flight is to build a flying machine. The seminal experiment by the American chemist Stanley Lloyd Miller in the mid-1950s showed that amino acids—the basic components of life—can be formed from a broth of lifeless chemicals when exposed to a flux of energy.

A good deal of the theoretical work in the sciences is also one of construction. This relates to the development of general models, frameworks, or theories that can accommodate diverse empirical observations.

Analytic studies in engineering are often dedicated to the determination of what is possible. For example, the hypothesis that manned spacecraft can visit Mars and return to Earth can be confirmed from our knowledge of the chemical properties of propellants, the mechanical properties of materials, and the physics of interplanetary flight.

Impossibility

A negative result, if it can be proven, is as useful as a positive result. In fact, most of the major scientific advances in the 20th century are of the negative kind: Einstein's Theory of Relativity says that there are no absolutes; Heisenberg's Uncertainty Principle states that position and momentum cannot both be determined simultaneously with arbitrary precision; Godel's Incompleteness Theorem says that there is no decisive algorithm to prove invalidity in predicate logic; according to Arrow's Impossibility Theorem, a particular set of reasonable assumptions will admit no consistent economic welfare function. These discoveries of the impossible actually serve to define the limits of the possible.

Trade-offs

An important class of results relates to interdimensional trade-offs. These may relate to the relationships between performance and efficiency, or time versus space, among others.

The student of economics quickly learns that life involves the pursuit of happiness under resource constraints. A simplified economy may have sufficient resources to produce each year exactly one of the following baskets of goods: 5 million muffins; or 800 videos; or 2 million muffins and 400 videos. These three alternatives define the "production possibility frontier" for the economy. The actual choice among the three alternatives will depend on the collective disposition of the consumers and producers. The nature of the choice is largely a subjective matter; but the trade-offs between muffins and videos is an objective phenomenon whose understanding facilitates the subjective decision. Happiness may be a subjective subject, but its pursuit can be supported by rational decision making.

In the realm of computer science, the area known as complexity theory deals with the consequences of differing algorithms for computational efficiency, and the trade-offs between requirements for memory versus computational time in solving a specific problem. These results support the design of computer systems, just as production possibility frontiers assist in formulating economic policy.

Nature of the Categories

The classification of scientific results into a set of categories provides a convenient framework for exposition and discussion. However, the various categories are not intended to be independent or mutually exclusive.

For example, the Theory of Relativity stipulates the lack of absolute frames of reference. This result may be classified as an example of the *impossibility* of determining an absolute reference as well as one of *invariance* of physical laws across reference frames.

Moreover, a result that falls into one category may engender results in other categories. As discussed previously, the laws of thermodynamics represent a *unifying* structure. The first law, however, stipulates the *invariance* of the total amount of energy in any insulated system. In addition, the second law can be used to deduce the *impossibility* of a perpetual motion machine.

The four categories of results highlight the different types of contributions. This framework can help to promote creativity by providing a platform for orienting research efforts and outlining the nature of results that may be sought.

Appendix B

Creativity in Humor

Analyzing humor is like dissecting a frog. Few people are interested and the
frog dies of it.   Elwyn Brooks White

A difficult problem, by definition, admits no obvious solutions. Hence the solution must be found in—or at least inspired from—unexpected quarters. The juxtaposition of incongruous elements has also been identified as the essence of humor.

Humor is a common attribute of creative persons. A study of school children, for example, cites humor and playfulness as distinguishing characteristics of creative individuals. This appendix delves into the various theories of the nature of humor, comedy, and related topics.

Introduction

We laugh in response to a diverse range of stimuli, known variously as wit, humor, comedy, or jokes. The dictionary defines humor as the "quality which appeals to a sense of the ludicrous or absurdly incongruous." In contrast, comedy is described as "a ludicrous or farcical event," while wit is characterized by "verbal felicity or ingenuity and swift perception, especially of the incongruous." A joke is defined simply as "the humorous or ridiculous element in something."

These definitions, while self-contained in themselves, do not readily distinguish the various stimuli. For proper differentiation, we must refer to the connotations embodied in these words. We may say that humor is a general term referring to any event that provokes mirth or laughter. A humorous comment is usually a light-hearted affair devoid of bitterness. A joke is a humorous narrative or, as in the case of the practical joke, a contrived event. Wit, on the other hand, refers to a humorous remark having a ring of truth. In our society, comedy often implies a professional production of humorous escapades such as those found in vaudeville or comic drama.

These connotations provide a convenient basis for differentiating humor in its various manifestations. From a historical perspective, however, the terminology did not always carry its modern connotations. We will later see how philosophers and other students of humor have interpreted these concepts in the past.

Laughing and smiling are curious phenomena that involve both voluntary action and spontaneous reflex. As a rule, physical reflexes, such as the involuntary removal of a finger from a lighted match, are simple responses with obvious survival value. On the other hand, laughter involves the contraction of facial muscles coupled with a peculiar ejection of air: activities that apparently serve no purpose in the quest for survival. Since laughter serves no known biological end, it has been called a luxury reflex whose purpose is to release tension.

A sense of humor grows over the course of a lifetime. The evolutionary phases are defined by a series of difficulties or anxieties encountered during the course of individual growth; the resolution of conflicts at each stage leads to the maturation of humor. The smile shows up in a human infant soon after birth, then develops gradually into the chuckle and the laughter of the adult.

Historical Development

Poets and philosophers throughout the ages have pondered theories of laughter and humor. In the fifth century B.C., humor first became a subject for philosophy and drama. Plato believed that laughter results primarily from an enjoyment of observing the weaknesses or misfortunes of others. He recognized ridicule and the ability to laugh at oneself as other sources of comedy.

Aristotle agreed with Plato's theory of ridicule, and confirmed the abusive character of jokes. However, Aristotle also introduced the notion of surprise. Surprise may be due to some disappointed expectation in the listener, or to the identification of a relationship between objects with little obvious connection. Aristotle also believed that laughter springs from an innocuous type of ugliness. An example is the comic mask which, although ugly and distorted, is not so grotesque as to evoke strong feelings of fright or repulsion.

The Roman orator Marcus Tullius Cicero believed that risibility springs primarily from unseemly things that are uncouth or ugly. Often the best way to make a point is through ridicule, although the subject matter must not be one the listener holds dear or regards as excessively painful.

In humor as in other domains, the Greek philosophers fueled lively debate among thinkers of the Renaissance. Since laughter often expresses a mixture of sadness and levity, the Western Europeans theorized that its source lies in the heart rather than in the brain.

The French mathematician and philosopher René Descartes argued that laughter can only result from two causes: hatred or surprise. Sometimes these ingredients are intertwined, as when an observer bursts into laughter upon noticing a scornful defect in another person.

The English philosopher Thomas Hobbes resurrected the Greek theory of derision. In *Leviathan*, published in 1651, he asserted that laughter springs from self-satisfaction. A person who notices a deformity or imperfection in another individual breaks into a scornful fit of laughter.

In the play *Tartuffe*, first performed in 1664, the French dramatist Molière highlighted the power of ridicule. The playwright opined that people can bear to be criticized outright, but not mocked or ridiculed. The moral aim of comedy, in fact, is to correct societal faults through amusement.

Another English philosopher, John Locke, established a parallel between wit and judgment. He defined wit as a collection of ideas whose mutual resemblance yields pleasant images. Judgment, on the other hand, involves differentiation of related ideas; one must seek out and highlight the differences to avoid being misled by superficial similarities.

The English lexicographer Samuel Johnson asserted that "every dramatic composition which raises mirth, is comick." In breaking away from some earlier theorists, he maintained that belittlement is not a necessary component of laughter. Moreover, the subject of humor need not be base or corrupt, nor the events trivial or fictitious.

The French writer Voltaire offered a simple view of laughter. In 1764 he asserted that laughter results from joy rather than contempt or vanity. The German philosopher Immanuel Kant proposed a mental and physiological view of humor in 1790. He identified laughter with the abrupt change in physiological processes when a tense expectation suddenly dissolves into nothing. The mix of sensations generates a feeling of health and is therefore pleasurable.

In 1876, the American essayist Ralph Waldo Emerson declared that our species is the only one capable of joking. He proposed that all comedy and jokes result from a deliberate dilemma, on the one hand making loud claims of offering the observer something, then later dashing his expectations. The break in a line of thought and the frustration of expectations leads to comedy, which manifests itself physically in the pleasant convulsions of laughter.

One of the most celebrated works on laughter belongs to the French philosopher Henri Bergson. The first point about the comic is that it relates to something human, whether in the form of a direct comparison or an indirect allusion. A river or mountain may be ugly or beautiful, but not comic. On the other hand, a dog or robot might be funny because of some resemblance to human looks or behavior. Conversely, human gestures that resemble machine-like actions engender laughter.

A necessary condition for humor is indifference or at least the lack of strong emotions. This perspective is illustrated in comic drama, when the misfortunes of a character appear risible to the detached audience.

On the other hand, laughter is a social phenomenon that calls for a feeling of kinship with other spectators, whether real or imagined. A newcomer in the midst of strangers will not share their levity. In fact, joking and laughter may cause him to become uncomfortable, until the newcomer identifies more with his companions and joins their merry-making.

A common source of laughter is the discrepancy between human intention and the state of the world. This happens when a person steps on a banana peel without changing his stride, then falls. It also occurs in a practical joke when someone tries to leave a room and runs into a closet instead. In a similar way, the entranced lovers who continue to whisper sweet nothings long into the night, oblivious to the closing of the cafe, provoke a humorous response in the observer. Humor also arises when a person continues to wear yesteryear's style of dress despite changing tastes.

As these examples illustrate, laughter results from inflexible responses to situations that call for versatility. It therefore serves as a corrective mechanism to promote adaptation, both in terms of adjustment to changing environments or conformity to societal norms.

Sigmund Freud advanced a systematic theory of levity based on psychological insights. In 1905 the founder of psychoanalysis theorized that the pleasure in jokes springs from an "economy in expenditure upon inhibition," the allure of the comic from an "economy in expenditure upon ideation," and the attraction of humor in "an economy in expenditure upon feeling."

Freud identified two types of jokes or witticisms: the "innocent" joke whose sole purpose lies in the enjoyment of a novel structure or concept, and the "tendentious" joke which serves as a vehicle for venting repressed feelings such as hostility or sexual desire. An example of an innocent witticism lies in the simplicity of the statement, "Time is nature's way of keeping everything from happening at once," or the peculiar reasoning in the remark, "If I had known I was going to live this long, I would have taken better care of myself." A tendentious remark is found in Henry Kissinger's observation, "University politics are vicious precisely because the stakes are so small."

For Freud, comedy is a superset of jokes. A common source of comedy is a sense of superiority over other people or objects. This occurs in practical jokes, verbal parodies, and visual caricatures. Comedy springs from the structure of an external event, in contrast to a joke, whose attractiveness springs from aired feelings. In other words, a joke is a comic event that originates from the unconscious. For this reason, an observer may find comic pleasure in solitude whereas a joke must be shared with another person to be enjoyed.

Jokes resemble dreams in that both are created in the realm of the unconscious, a netherland free of social constraints. Here, concepts may jostle freely and coalesce to form novel structures. Bypassing the censoring mechanisms of rationality and propriety allows for the genesis of new ideas that otherwise could not have been conceived.

An excess of distressing emotions interferes with comic relief. When wanton activity causes excessive damage or inflicts pain, the pleasure in comedy or wit vanishes. In contrast, humor arises from a release of uncomfortable feelings. More specifically, humor occurs when an observer is led to develop a sense of discomfiture, then suddenly realizes that this feeling is unnecessary. The discomfort is then discharged through laughter. In short, discomfiture is the source of humor but an impediment to wit and comedy.

Freud regarded humor as the highest of the defense mechanisms. In ordinary psychic processes, uncomfortable feelings are repressed into the subconscious in a futile effort to banish the discomfort by ignoring it, only to have it resurface at a later time. In humor, on the other hand, distressing emotions are faced squarely and transformed into pleasure through discharge.

The American editor Max Forrester Eastman declared that humor has both negative and positive aspects. The negative aspect is related to the disappointment of a playful interest in the listener. The positive or active theme involves the engagement of some emotional interest, such as rightful indignation, repressed sexuality, or affinity for truth. An example of how these two aspects coexist in a single joke lies in the epigram, "Marriage is one good way to kill a romance." The listener, on hearing the first part of the sentence, expects perhaps some nicety such as, ". . . to lead a happy life." This playful interest is disappointed when she hears something quite different from that anticipated. The positive aspect of this quip arises from its satisfying, to some extent, our love for truth—it is not uncommon for the routine of marriage to displace the excitement of romance.

After studying different kinds of humor, ranging from practical jokes to poetic levity, Eastman offered several principles for jokes:

- The listener must be genuinely interested in the subject: humor involves feelings rather than merely intellectual curiosity.

- The emotions aroused in the listener must not be too strong or painful.

- The elements of the joke must appear to be naturally induced rather than forced or contrived.

- A joke must occur in a flash, rather than be drawn-out or explained.

- Style and substance must be compatible in a joke: practical jokes should not be poetically told, nor subtle jokes practically told.

- A joke must be novel enough to surprise the listener.

- The positive aspects of a joke must outweigh the negative.

Throughout recent history, from the classics to the moderns, people of various origins have reflected on the concept of humor and laughter. These theorists reinforced each other on some ideas, and contradicted each other on others, but each drew on existing thought and contributed to the overall theory. The process has continued to this day to leave us with the latest theories—theories nevertheless still subject to evolution.

Modern Theories

Modern theories of humor involve two interacting facets relating to intellectual and emotional structures. A discussion of intellectual structure involves general attributes of humor as well as specific techniques. The emotional component involves a modicum of aggression or apprehension: not too much, just enough to add spice to the experience.

Intellectual Structure

Fusion. Routine thinking, such as deductive logic, occurs within a single field of association; but creative thinking, such as the formulation of a joke, involves two or more planes of thought. The term "fusion" or "bisociation" is used to denote such a meshing of two planes. A joke, for example, couples two ideas that resemble, contrast, or otherwise relate to one another. The listener may be led along one pathway, then suddenly exposed to another interpretation.

General Requirements. The essential ingredients of humor are surprise and economy. Freshness and originality lend strength to a joke by enhancing the element of *surprise*. As in other realms of creativity, however, originality is hard to come by. In the commercialization and mass-production of humor, the creation of tension through repetition and suspense has often replaced the more effective tactic of originality as a way of injecting surprise. This technique is used when fifteen clowns emerge from a tiny Volkswagon, or when an endless stream of water appears to flow from a severed hose.

With increasing sophistication in humor comes a greater reliance on *economy*. Here, economy refers to the use of implicit suggestion instead of explicit itemization. Mark Twain once said, "Always do right. This will gratify some people and astonish the rest." This comment highlights the implicit assumption that people choose wrong over right as a matter of course. The humor occurs in the fact that the message is conveyed indirectly, yet more strongly than by direct declaration.

Particular Techniques. Many techniques are available for eliciting laughter from a listener. Some of the most popular methods relate to brevity, exaggeration, unification, double meaning, contrast, and parallel structure.

Brevity. Levity is a child of brevity. A celebrated quotation on humor comes from William Shakespeare:

> Therefore, since brevity is the soul of wit
> And tediousness the limbs and outward flourishes
> I will be brief.

Brevity is a necessary but insufficient ingredient for a joke; otherwise every terse comment would be one. More specifically, a joke conveys its message in words which are too few by the standards of logic or everyday convention. Sometimes a joke can say something without actually saying it.

An example is found in the comment "Victory finds a hundred fathers, but defeat is an orphan." This brief sentence speaks volumes about the human tendency to claim personal association with an achievement and deny ties to a failure.

Another example is found in Woody Allen's plea: "More than any time in history mankind faces a crossroads. One path leads to despair and utter hopelessness, the other to total extinction. Let us pray that we have the wisdom to choose correctly." On the surface, this quote declares that we face a dilemma of two alternatives, each equally morbid. But the implict message is that we do, in fact, have a third choice: pursue a course of action that steers clear of both these paths.

Exaggeration. Exaggeration, whether in the form of hyperbole or myopia, can lead to picturesque images. An example of an understatement lies in Jane Austen's remark, "What dreadful hot weather we have! It keeps me in a continual state of inelegance."

In response to inaccurate reporting by some newspapermen, Mark Twain once protested, "The reports of my death are greatly exaggerated." The humor springs from the understatement in the phrase "greatly exaggerated," since any unfounded report of a death is not merely exaggerated but categorically false.

A popular form of exaggeration is simplification or irrational emphasis. Simplification may take the form of specific attributes, as exemplified by caricatures of drunken Irishmen or aristocratic Englishmen. An example of emphasis occurs in an observation about the sexes. In a play by the Irish writer Oscar Wilde, one of the characters declares:

> Men always want to be a woman's first love. That is their clumsy vanity. We women have a more subtle instinct about things. What we like is to be a man's last romance.

Humor springs from the simplified aims of the two sexes in conjunction with the contrast between the first and last affairs.

Unification. Sometimes ideas can be related by definitions and mutual references to one another or to a thing. The similarity of concepts, reinforced by the similarity of words, ties the entire package into a coherent whole. A unifying perspective of the human experience is found in the epigram, "In youth, we long for maturity; and in maturity, youth." Another example in a similar vein is, "Some day we'll look back on these miserable times as the good old days."

Unification is often achieved simply by linking ideas together with the use of the conjunctive "and." When ideas are held together in this way, the listener readily assumes them to be of equal magnitude in the hierarchy of mental associations. Any deliberate incongruity in such an implication gives rise to risibility. An example lies in the motto "For God, country, and Yale." The humor arises from the implication that the status of Yale University is comparable to those of God and country, a view to which some would take exception.

Double Meaning. The class of double-meaning jokes can be loosely divided into three categories. The first relates to the double meaning of a word that denotes a name and a thing. This happens when Frank turns to his date Melody in a discotheque and declares, "To be frank, I can't stand this melody." Each of the two words, *frank* and *melody*, alludes to both a person and a concept or object.

The second category involves double meaning due to literal and metaphorical senses. An example lies in Mark Twain's remark, "Familiarity breeds contempt—and children." The first part of the sentence invokes a metaphorical interpretation of the verb *breed*, while the latter engenders a literal perspective.

The third category relates to literal double meaning. This is the purest form of multiple-usage. The double meaning springs from the various literal meanings of a single word. Double meaning is found in the remark, "The difference between Los Angeles and yogurt is that yogurt has an active, living culture." The last word in the sentence—*culture*—refers both to a bacterial colony and the trappings of civilization.

Consider the following comment: "Husbands are like fires. They go out if unattended." Humor occurs when the phrase "go out" is interpreted in two senses, as in the expiration of a fire and a sortie in search of amusement.

Contrast. The confrontation of contrasting factors or the representation of an idea by its opposite can lead to incongruities that elicit laughter. Derek Bok of Harvard University illustrated the technique of confrontation when he quipped, "If you think education is expensive, try ignorance." The representation by an opposing image also occurs in the portrayal of an old man by a child in comic theater.

Parallel Structure. Humor can be created by taking certain verbal material and changing its arrangement. The technical quality of the joke varies inversely with the amount of arrangement required. An example of an effective structure is found in the epigram, "Success is getting what you want; happiness is wanting what you get."

There are many other techniques of humor, including absurdity and faulty thinking; but, unfortunately, we do not have the space to discuss all of them here.

Emotional Structure

Humor and other forms of creative expression have no clear-cut boundaries. The spectrum of humor—from the practical joke to the brainteaser—shows a gradual change in the degree of emotional intensity. Humor that relies on an emotional response contains at least a trace of aggression or apprehension.

Aggression. The aggressive element may appear as contempt, malice, or merely a lack of sympathy for the victim of the joke. The coarse laughter associated with a practical joke is simply muted aggression. The morbid joke arises from sadism that is denied full expression, and satire springs from a feeling of self-righteousness. Children are amused by ridiculing others, but cannot understand witty comments.

As discussed previously, aggression as a fundamental ingredient in humor has been recognized by theorists throughout the ages. According to Aristotle, laughter results from deformity and surprise. Descartes believed that laughter springs from joy coupled with surprise, hatred, or both. Henri Bergson identified laughter as a corrective tool to discourage abnormal behavior.

The belittlement of an enemy through a joke, parody, or satire leads to a psychic victory. Since a joke allows one to remain within social restrictions, it is possible to disparage an enemy and still retain a pristine conscience. An example is found in the story of an emperor touring his provinces. One day the emperor noticed a man in the crowd who looked remarkably similar to himself. "Was your mother ever in service in my Palace?" asked the emperor. "No, your Highness," replied the other, "but my father was."

Since an emperor cannot be challenged at will, the commoner offered a repartee that served to clear his mother's name and in the process questioned the emperor's own legitimacy. Here, the joke is a psychic weapon.

An anecdote provides us with another illustration of how aggression can be humorous. One day, as Oscar Wilde was engaged in conversation, an enthusiastic gentleman approached the writer, slapped him on the back, and exclaimed, "Why, Oscar, you are getting fatter and fatter!" The listener immediately empathizes with the victim of this unprovoked insult. Without turning around, Wilde retorted, "And you are getting ruder and ruder." This punchline elicits laughter because the listener feels that the ill-mannered acquaintance has been most deservedly—and humorously—upbraided.

Apprehension. Not only can laughter spring from aggression but also from apprehension. This is illustrated by the change in a child's state of mind when strange, growling noises in another room turn out to be those made by his sibling.

Apprehension as a source of laughter has been identified by many theorists. When a comedian tells a story, he deliberately creates tension in his listeners. This tension builds up but is never allowed to reach its expected climax. The only recourse is to discharge it through laughter.

Laughter involves a clumsy physical response. If its purpose is to relieve tension, where does the energy come from? And how is it that even a small intellectual stimulus can provoke such a physical response? The paradox can be resolved by regarding laughter as the result of catalytic action. A mental stimulus, large or small, triggers the discharge of accumulated emotion and tension. The stored energy can spring from numerous sources, ranging from latent fears to suppressed aggression and sexuality.

Interplay of Thought and Emotion

The currents of reason may change direction within a split second, but emotions cannot. According to the physiological model of laughter, a person laughs because her emotions and feelings have more inertia than her thoughts and ideas. The difference in momentum has a physiological basis. Emotions rely on the sympathetic nervous system and the hormones that act on the entire body through mechanical and chemical pathways. These processes require several seconds or even minutes to take effect. On the other hand, intellectual processes occur in the neocortex of the brain and can transpire within a split second.

In infrahuman species, thinking may be bound inextricably to feeling. Through the course of evolution, humans have acquired the ability to reason largely independently of their emotions. However, the separation is only partial, since thought and feeling can enhance or interfere with each other.

A substitute for aggression is laughter. Like a fierce game of tennis or other aggressive activities, laughter provides a vehicle for emotional discharge.

Comedy is not an absolute quality of nature. It depends on the viewer, and as such the same event may be comic to one observer but tragic to another. The fine line separating comedy and tragedy depends on an individual's level of emotional investment.

Strong emotions interfere with the reception of a comic event. Freud believed that the comic situation palls as soon as the action causes damage or inflicts pain. This is certainly true for the person caught in a joke and forced to serve as the victim.

Freud identified the following conditions as ingredients of the comic situation.

- *Cheerful mood.* In an atmosphere of contagious mirth, everything seems comic. The object of jokes and comedy is, in fact, the attainment of this euphoric state.

- *Expectation of the comic.* An intention to make something appear comic is all that is required to set the wheels of mirth and laughter in motion.

- *Lack of serious mental activity.* Deep concentration or intellectual work interferes with the ability for catharsis. Comedy can arise only in the absence of intense mentation or abstract reflection.

- *No attempt at evaluation.* Something cannot be comic if one's only interest is to compare it with a set standard. A student's mistakes during an oral test may evoke laughter from his friends, but his behavior is unlikely to have the same effect on the examining professor.

- *Weak interest.* The appreciation for the comic is enhanced by indifference or lack of strong feelings and interests.

The need for detachment in a comic situation is illustrated by a type of humor known in Freudian literature as "gallows humor." An example is the story of a condemned man being led outdoors for execution early one morning. He looks up to the clear, blue sky and cheerfully observes, "Well, the day's beginning nicely." On hearing this story, the listener first finds herself filled with pity for the fate of the condemned. But the pent-up emotion is discharged in mirth, since the victim himself makes light of the situation. A similar case occurs in a story about Oscar Wilde. It has been said of the Irish writer that his dying words were, "This wallpaper is killing me—one of us has got to go." The pity that emerged on the subject's behalf becomes meaningless and is conveniently discharged through laughter.

The sense of humor varies from person to person, both in quantity and quality. Even so, wit and humor can be used to bridge intellectual disparities and emotional chasms by providing a common ground for mirth and release of excess energy. As with most human abilities, the capacity for humor is poorly developed at birth, evolving only slowly with physical maturation. The appreciation of humor is enhanced by a well-developed intellect; it cannot flourish from a skimpy knowledge base, nor in a vortex of passion.

The personality of the wit is reminiscent of those of both the sadist and the optimist. The wit is quick and deliberate, often using his intellect to reduce his victims with efficiency and dispatch. But life is too short for tears, and the humorist prefers to highlight the springs of joy. He attempts to smooth out the vicissitudes of life through veiled aggression, fantasy, and other tactics.

Summary

Writers throughout the ages have recognized humor as a rich arena for creativity, perhaps even the quintessence of intelligent behavior. Like other creative pursuits, humor involves the principle of multidistance (as explained in Chapter 2): some attributes of the comic situation are close in conceptual space in that they relate to a common theme, while others are distant.

The factors of creativity permeate all forms of humor. The *purpose* of a humorous situation is to elicit some emotional response in the audience, a response that may range from a chuckle to boisterous laughter. The *diversity* factor in humor involves the fusion of two or more objects not normally associated with each other. The fusion involves the identification of *relationships* that tie the objects through a unifying theme. Many forms of humor evoke striking *images* ranging from the picture of a child wearing an oversized coat to the caricature of national stereotypes. Unless the comic situation is to remain solely in the mind of its orginator, it must be *externalized* in some form. The medium of delivery often contributes dramatically to the power of the humorous situation, whether expressed in the form of words, imagery, or physical structures.

Appendix C

The Memory Agent

The primordial images of the collective unconscious . . . are deposits of thousands of years of experience of the struggle for existence and for adaptation.   Carl Gustav Jung

Creative problem solving involves the integration of diverse elements. An essential component in this process is human memory, both as a store of knowledge and as a work area. The contents of archival memory facilitate the generation of novel elements relating to a solution, while working memory is a board on which to craft a solution.

This appendix discusses a number of interesting observations relating to memory and creative performance. First is a discussion of the impact of experience on achievement, followed by a model of memory and implications for its enhancement.

Age and Achievement

The amount of knowledge required for creative work depends on the field. One study of 738 scholars, scientists, and artists born after 1600 indicates that artists often reach their peak of productive output earlier than scientists, who in turn precede scholars such as historians and philosophers. Artists tend to reach their zenith in their 40's, while historians and philosophers do not peak until their 60's. The golden age for technical contributors spans the entire spectrum, from the 40's for biologists to 50's for geologists and 60's for inventors. Mathematicians defy categorization: they are as productive in their 30's as they are in their 60's. Another study indicates that in many areas of intellectual achievement, the most creative period occurs between the ages of 30 and 45.

Another study focusing on the achievements of 420 literary contributors shows that the modal productive age for the most renowned works of poets is 38.5 years; the figure increases for writers of imaginative prose to 42.6 years, and again for writers of nonfiction to 50.3 years. A related study examined the average age at which different types of writing were produced. For lyric poetry the age is 33.9 years, for novels about 42.1, and for informative prose around 46.7.

An interesting observation is that the average life span of poets is six years less than that of prose writers. However suggestive the data, we should refrain from concluding that poetry is hazardous to one's health. A more plausible explanation is that poems tend to be short, imaginative pieces that may be completed in minutes or months. On the other hand, prose often finds expression in books which usually take much longer to complete. Further, informative prose often requires the accumulation of knowledge over time, more than is the case for many fictional themes.

In modern times, our collective knowledge of the natural world has grown at an exponential rate, as has the complexity of our social structures. It should not surprise us, then, to learn that the amount of knowledge an individual needs to succeed in his work has steadily increased in a number of fields. We may infer from this that the requisite knowledge for producing a worthwhile piece has been increasing over time.

However, a study of achievers in intellectual fields such as literature, philosophy, science, and invention indicates that the age of greatest contribution has been decreasing. In almost every field, the workers born between 1775 and 1850 exhibited their ingenuity at an earlier age than their predecessors. In fields such as geology, the most productive age has declined from an average of 45 years to 35 years. One possible explanation for the decline is that our collective store of knowledge is better codified and transmitted more efficiently to new generations of students.

Although people appear to be making their greatest contributions at an earlier age, the time required to attain recognition seems to have increased. For example, the median age of workers at the time of their election to the U.S. National Academy of Sciences rose from 41.3 years in the late 1800s to 51.8 years by World War II. Similarly, the average age of political leaders has been increasing steadily. For example, the median age of cabinet members in the United States has increased by 13 years over the course of 150 years, from 46.9 years of age in the early 1800s to 60.1 by World War II.

Empirical studies indicate that creativity increases with age, peaks toward the middle years, then declines according to a curve that is remarkably independent of external factors. Dean Keith Simonton has proposed a two-component model to depict this phenomenon. First an assumption is made that the potential for creative output is fixed for all time at an early age for a given individual. Further, the rate at which creative tasks are addressed is deemed to be proportional to the creative potential at any moment, and the rate at which solutions are finalized is proportional to the number of tasks in progress. This model can be transformed into a set of equations whose solution takes the general shape shown in Figure C.1.

The general form of the curve models the observed data remarkably well. Even so, the assumptions underlying the model may be simplistic. For example, the reason for the declining curve might well lie in causes other than a fixed capacity for creative work. Surely it may be due, at least in part, to an entrenchment in tried-and-true methods, a tacit acceptance of the renowned maxim "If it ain't broke, why fix it?" Or it may be caused by the distractions of success, such as the directorship of a laboratory, service on governmental panels, and correspondence with a widening circle of acquaintances. The rewards of success can thereby serve to shackle creativity.

According to the entrenchment hypothesis, the peaking of creativity in middle age may be explained in an alternative way as the product of two opposing factors. The amount of knowledge relevant to a specific problem increases with the age of an individual, but the likelihood of introducing a fresh perspective tends to decrease. The quality of the resulting solution depends on the product of these two factors. It therefore increases initially, reaches a peak, and begins to decline with age as shown in Figure C.2.

Despite obvious similarities in the reservoir and entrenchment models, the latter has operational implications. Since creative capacity is fixed according to the reservoir model, a prolific creator—should she live long enough—faces a future of emptiness. On the other hand, the entrenchment hypothesis supports a more optimistic outlook in which the loss of novelty is a fluid process rather than the consequence of a depletionary process. According to this model, a decline in creativity is not a necessary consequence of a productive life. The argument is supported by powerhouses such as Picasso who generate major contributions until their final days.

The trick is to maintain an open mind, itself a hallmark of creative personalities. The

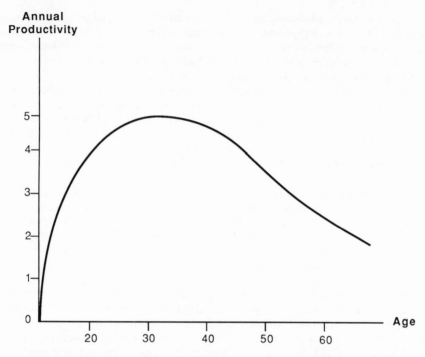

Figure C.1. Creative productivity as a function of age. [After Simonton (1983).]

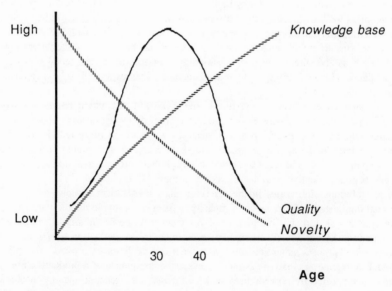

Figure C.2. Quality of ideas as the product of fresh perspective (novelty) and relevant information (knowledge base).

mind and body do not always keep pace in our journey through life. Some of us slip into an uncaring bliss at an early age, while others remain forever young. The luckiest among us maintain a childlike curiosity despite the vicissitudes and routine of life, the former squelching the spirit and the latter the mind.

The medical community tells us that physical well-being is largely a state of mind. Creativity, as the consequence of high-level thought, must be the quintessence of the notion of child-of-mind. The tendency for creativity to decline with maturation can therefore be arrested, and even reversed, simply through conscious decision and deliberate action.

A large knowledge base is an important ingredient of the creative process. The creative synthesizer requires a rich base of techniques, information, and facts to concoct an original solution to a problem.

Memory is the mediating agent that relates a rich knowledge base to the diversity that characterizes the solution to any difficult problem. Fortunately, memory can be trained. The mechanism of memory exhibits certain characteristics that may be explicitly utilized to enhance the operations of encoding and retrieval.

Two-Stage Memory Model

Human memory may be partitioned into two major components: short-term and long-term storage, as shown in Figure C.3. Once an image is perceived, it enters into short-term memory, a temporary storage buffer. The image will not persist for more than a few seconds, although it is possible to extend this period deliberately; for example, a telephone number may be retained for longer periods by repeating it aloud every few seconds. The capacity of short-term memory is also very limited, for it can hold no more than 5 to 9 items at a time.

Items in the short-term buffer then enter long-term memory, a permanent storage facility. This facility has an infinite capacity for practical purposes. Its contents also fade slowly; the probability of retrieving a specific item decreases over time, but it is possible that the memory never vanishes completely.

In other words, the limitation in long-term memory seems to be one of retrieval rather than encoding. This premise is supported by studies in hypnosis. For example, the witness to

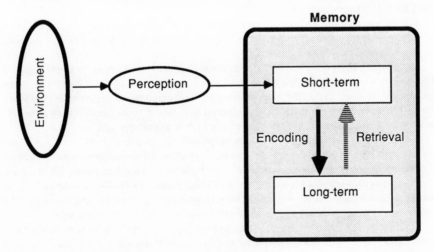

Figure C.3. Two-stage model of human memory.

a crime may say that she does not remember the license number of the getaway car. Later, however, she is able to recall the number after entering into a hypnotic trance.

Evidence such as this suggests that whatever is perceived finds its way into long-term memory. However, the likelihood of retrieval increases with the level of deliberation in recording the image or item of knowledge.

Interference

Interference refers to the coupling effects of interacting components. The coupling may be of two types, positive or negative.

In *positive interference*, knowledge of one component aids in the encoding and/or retrieval of another. Various objects are linked with each other through a network of associations. For example, the knowledge that the far side of the moon is never visible from the Earth is reinforced by the knowledge that the period of rotation about its own axis is equal to the period of revolution around the Earth.

In a similar way, knowledge that market share tends to increase the profitability of a company is reinforced by an understanding of the causative factors. More specifically, the larger revenues allow for economies of scale in operations, plus the amortization of development costs over a larger revenue base, as well as the rapid descent along the learning curve as cumulative production decreases unit costs.

Hence a prescriptive implication of positive interference is the deliberate fusion of new knowledge with old information through a rich network of associations. In other words, new knowledge to be "memorized" should be linked to existing facts in memory. The likelihood of future retrieval increases with the number of links and the strength of each bond.

In contrast, *negative interference* refers to destructive interaction among components: knowledge of one item hinders the retrieval of another. Destructive interference may further be classified into proactive, retroactive and bilateral types.

Proactive interference occurs when existing knowledge hinders the acquisition of new knowledge. For example, the conventional typewriter based on the QWERTY keyboard, was designed deliberately to impede typing speed: in the early days of the typewriter, the mechanical linkages would jam when operated rapidly. Although more efficient key configurations have been proposed, people cannot adapt readily to a new configuration and the cumbersome QWERTY still reigns supreme.

For the same reason, a beginning language student would be advised to separate the study of French and Spanish by placing Art or Physics between them, instead of studying one language immediately after the other.

Retroactive interference refers to the impairment of previous learning by newly acquired knowledge. As a simple illustration, consider the above student who is taking both French and Spanish lessons. In the first hour, she may have mastered the conjugations of the French verb *être*. If this period is immediately followed by a lesson in the Spanish verb *estar*, the student may later discover that her competence in *être* has disappeared.

The detrimental effect of retroactive interference is depicted in Figure C.4, which shows the effect of time on retention rate for nonsense syllables. These are monosyllabic "words" without meaning, such as NAR and GOW. When the subject remains awake following the memorization of a set of such words, the recall rate drops to 50% after 1 hour and almost vanishes after 8 hours. If the subject falls asleep immediately after learning the words, the rate of loss is much smaller, and in fact tends towards a higher asymptotic value.

It should be noted that the retention rate declines precipitously for items without meaning, such as nonsense syllables or telephone numbers. When the memorized item has more meaningful features, the decay over time is more gentle.

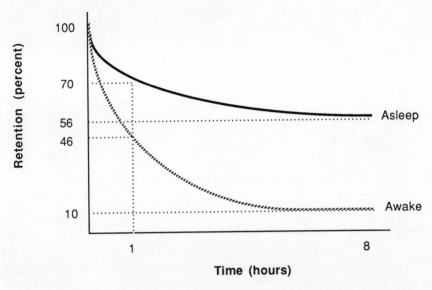

Figure C.4. Retention of nonsense syllables as a function of time, when learning is followed by wakefulness versus sleep. [After Jenkins and Dallenbach (1924).]

The retention rate is highest for conceptual items of knowledge. Hence the knowledge that molecular vibrations vanish as the temperature approaches absolute zero is more likely to endure than the information that absolute zero occurs at -273.16 degrees Celsius.

The prescriptive implication is that "nonsense" material should be learned immediately before a quiescent state such as resting or sleeping. The conceptual material will more likely persist in memory even if it is followed by some other engaging activity that requires concentration.

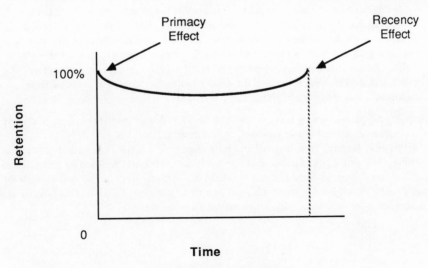

Figure C.5. Primacy and recency effects on learning retention.

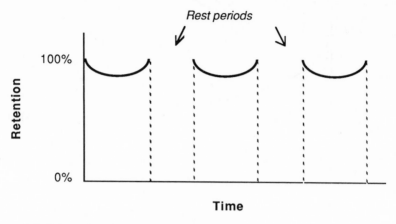

Figure C.6. Effect of spaced learning on retention.

A third type of negative effect on learning is that of bilateral interference. This consists of the primacy and recency effects in which material learned at the beginning or end of a session is more likely to be recalled than that learned in the middle (see Figure C.5). This condition may be considered to result from proactive interference, in which the initial learning impedes subsequent learning, and retroactive interference, in which the final material overshadows the intermediate learning.

A prescriptive implication of bilateral interference is that of "spaced" learning, in which a given amount of material is learned in modules separated by other activities (see Figure C.6). The cumulation of primacy and recency effects contributes to the overall retention as compared to "massed learning" in which the same volume material is learned in fewer modules.

The American psychologist William James writes of the inefficiency of massed learning, such as cramming immediately before an exam.

> Things learned thus in a few hours, on one occasion, for one purpose, cannot possibly have formed many associations with other things in the mind. Their brain-processes are led into by few paths, and are relatively little liable to be awakened again . . . on the contrary, the same materials taken in gradually, day after day, recurring in different contexts, considered in various relations, associated with other external incidents, and repeatedly reflected on, grow into such a system, form such connections with the rest of the mind's fabric, lie open to so many paths of approach, that they remain permanent possessions.

The utility of spaced learning is consistent with the view that all mental processes, including the formation of associations in memory, require time to transpire.

In general, learning new material in small, digestible modules is more effective than ingesting it in a single gulp. On the other hand, it is also possible to partition the material into chunks so small that the administrative overhead of switching from task to task exceeds the benefits to be gained from spaced learning. The optimal duration of a module may run from a few minutes to several hours, depending on the nature of the task.

Appendix D

Creativity Support System

Imagination deserted by reason produces impossible monsters: united with her, she is the mother of the arts and the source of all their marvels.
Francisco de Goya y Lucientes

Creativity springs from the deliberate selection of options to satisfy a challenging objective. By encoding decision rules of creative thinkers in specific domains, computer systems can assist in generating solutions to difficult problems. For arenas of sufficiently limited scope, a sophisticated program may match—or even exceed—the performance of a human.

A comprehensive theory of difficult problems and creative solutions would facilitate the development of a knowledge-based system to support innovative thinking. Such a Creativity Support System (CSS) will draw on existing knowledge in the realms of cognitive psychology, artificial intelligence, computing hardware, and interface techniques. Some critical fields and technologies in these realms are given in Figure D.1.

The factors of creativity represent the ingredients for novel problem solving. A strategy for enhancing creativity, then, is to develop tools that promote or enhance these factors.

One dictionary defines an agent as "a means or instrument by which a guiding intelligence achieves a result." This characterization seems appropriate for the supporting agents of creativity: memory, imagery, and externalization. Each of these components may be enhanced by software tools.

These tools must accommodate a rich knowledge base of *objects* as well as the potential *relationships* among them. Since the potential space of relevant objects and their relationships is infinite in practical terms, techniques are needed to highlight the most relevant items. In particular, the objects and relations that seem most promising to the problem at hand must be identified.

This selection process may be viewed as deterministic or probabilistic. In the former perspective, items are classified categorically. At each stage of the deliberative process, certain objects or relations are deemed to be acceptable for further consideration, while others are not. In the probabilistic scheme, each item is regarded stochastically. Some items are more likely to be selected than others. Of course the deterministic case may be viewed as a special case of the probabilistic approach in which the likelihood values are binary. If an item is acceptable, its probability for further consideration is 1; otherwise the value is 0. An architecture for incorporating these concepts in software is the subject of the next section.

Software Architecture

The software architecture for a Creativity Support System is shown in Figure D.2. The proposed system is tailored after a system developed for more limited objectives in design

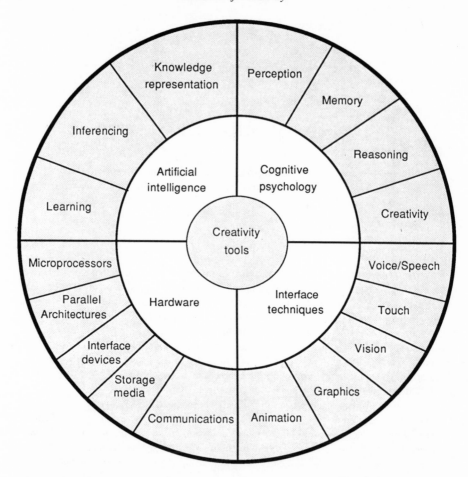

Figure D.1. Referent knowledge bases for developing creativity tools.

and manufacturing. The architecture consists of two primary modules, the Domain-Independent and Domain-Dependent Modules. The former component contains knowledge independent of the problem domain, while the latter contains information specific to the task at hand. These structures are described below in greater detail.

The Domain-Dependent Module contains a repertoire of the uses to which the CSS may be put. The system also has access to diverse fields or disciplines that may pertain to the problem, such as knowledge of statistical methods, financial accounting, materials properties, or others. Some of this knowledge may reside within the system, while other portions may be available in databases accessed through telecommunication networks. These knowledge bases reflect the need for diversity in addressing difficult problems.

The Case Base contains knowledge of a given problem, including specification of the task and the solution as it develops. For example, a software package for automotive design might specify an engine of certain power, subject to constraints on size, weight, cost, and fuel consumption. The solution to this particular design exercise is stored in the Case Base.

The Domain-Independent Module contains generic capabilities such as general-purpose strategies and representation techniques. Knowledge is represented and utilized through a set of linguistic and sensory vehicles. Declarative knowledge may be represented through mech-

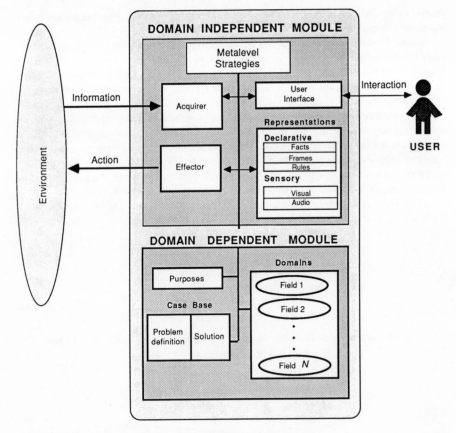

Figure D.2. Architecture for a Creativity Support System (CSS).

anisms such as facts, frames, or rules. These may serve as the foundation for linguistic constructs or images such as visual or auditory patterns.

These structures may be manipulated through operators such as inference schema for transforming knowledge from one state into another. The use of specific operators throughout the reasoning process is directed by metalevel strategies.

New knowledge is sifted and encoded through the acquisition module, which serves as an interface between the environment and the internal base of domain-specific knowledge. The need for externalization is served by two modules. First is the User Interface, which communicates with the human decision maker through audiovisual input and output. The second vehicle for externalization is the effector, which provides for a durable embodiment of the solution. The "hardcopy" may take the form of printed prose, drawings, mechanical prototypes, or other modes of externalization.

A number of these ideas have been implemented in the form of a prototype software package to assist in creating poems. To this program we now turn.

Poetry Generating System

The conception of creativity as the purposive fusion of disparate ideas has been implemented on software for generating simple poems. The Poetry Generating System was written in the programming language of Prolog by Mia Paget and Nancy Gilman as a student project at the Massachusetts Institute of Technology.

A central task in developing a poetry advisor is to encode literal as well as figurative and other relevant information. In a poem the connotations are of primary import and the denotations are secondary. Another type of knowledge involves heuristics for adding new words and phrases dynamically, as the poem develops. These types of knowledge constitute the core of expertise that must be encoded into software.

The connotations of a word lie at various distances in the conceptual space of the user. An operational way to define this distance is in terms of the amount of time required to invoke a connotation based on the original word. Synonyms are close, as are antonyms. The images conjured up by a word are more distant, as are analogies. For example, the synonym "joyful" is close to the key word "happy," while "sad" is farther away. The image of a smiling child is more remote, as is the analogy "happy relates to sad, as birth relates to death."

These types of knowledge are encoded in the computerized poet. The entry for the word "spring," for example, contains information such as the following:

Syllables	1
Parts of speech	noun, verb
Synonyms	grow, appear
Antonyms	winter, fall, die
Images	birds, flowers
Anologies	rebirth, alive, grow

The program must also contain knowledge of the structure of various types of verse. The format of the haiku, for instance, has the following constraints:

Total lines	3
Line 1	5 syllables
Line 2	7 syllables
Line 3	5 syllables

The work of the poetry advisor consists of three phases: initiation, construction, and editing. In the *initiation* phase, the user enters a list of themes such as "sunrise," "flowers," or "water." These themes serve as alternative centers of focus for the words to be added sequentially to the poem under construction.

The second phase of *construction* involves the actual selection of words to generate a candidate poem. This stage consists of two facets, deterministic and probabilistic constraints. The *deterministic* module specifies the nature of the subsequent word to be selected as a function of the type of poem, the parts of speech for the preceding words in the current line, and the number of syllables in that line.

All the available information is passed to the *probabilistic* module, which selects the next word subject to the deterministic constraints. This stage involves the random selection of a theme among those specified by the user in the initiation phase. The next step involves another random selection of the distance from the theme, in terms such as antonyms or images. The subsequent word is picked subject to these constraints. If no such word exists in the database, then a new iteration is attempted after some modification, such as the selection of an alternative theme. In the construction phase, new words are successively added to a line until the deterministic constraint module specifies that the line is complete. In a similar way, new lines are generated until the poem is finished.

Finally, the *editing* phase allows the user to modify the poem. This may be done directly, by replacing a word or an entire line; or indirectly, by asking the construction module to generate a new word or set of lines.

The Poetry Generating System, in interaction with the user, produced haikus such as the following:

> A petal springs then
> sings softly, however pale.
> Rebirth sprouts anew.
>
> Some pale pale flowers
> softly sing. Red rose and pale
> rose softly bring spring.
>
> Many new bursts seed
> gently, while the pale doves grow,
> the flowers burst new.

The user's contribution in the generative phase was limited to the interactive specification of parts of speech and to suggesting a new word or theme in the occasional instances when the construction module came up empty-handed. In addition, the user edited each poem in minor ways after it was generated. The editing was limited primarily to the use of punctuation, capitalization, and verb conjugation. (Grammatical functions such as verb conjugation are straightforward compared to the selection of key words and were not implemented in the prototype software.)

These poems in particular exhibit a subdued serenity. But the tenor of the verses will obviously depend on the selection of themes as well as the working vocabulary available to the program.

Conclusion

The Poetry Generating System was designed to demonstrate the notion that creativity lies in the coherent synthesis of seemingly unrelated ideas. The prototype software consists of fewer than 200 statements in Prolog, including the user interface, a primary vocabulary of about twenty words, and some procedures to assist in debugging the program.

Despite its brevity and simplicity, the system displayed a remarkable ability to generate interesting verse. The quality of generated verse may be improved by enlarging the working vocabulary, as well as by encoding the knowledge of expert poets. The resulting system could be used as an intelligent assistant in creating verse, much as a computer simulation program of a national economy can assist in exploring the consequences of alternative monetary policies.

This book has explored the proposition that creativity is not an entirely esoteric or mysterious phenomenon. Our enhanced understanding of difficult problems and creative solutions can lead to more systematic ways to address challenging tasks. It can also serve to empower computer systems to assist creative thinkers in developing solutions, and perhaps even in automating certain creative tasks that yet lie beyond the grasp of conventional software.

Notes

Chapter 1

P. 3   Quote by Chesterton: Chesterton in Dickens (1945), p. viii.

Chapter 2

P. 8   Quote by Drucker: Drucker (1985), p. 255.

P. 9   "Creativity is . . . result of attaining special goals": Perkins (1981), p. 101.
   Quote: "Creative thinking is not an extraordinary form of thinking . . . thinker produces it": Weisberg (1986), p. 69. Italics in original.
   ". . . make tangible something that he thinks or feels": Guilford (1977), p. 161.
   "Creativity implies novelty but not vice versa": Amabile (1983b), pp. 357–376.

P. 10   Discussion of algorithms vs. heuristics: The notion of algorithms and heuristics described here originates from the work of computer scientists. Some psychologists use the terms in the same way (see, for example, Hilgard and Bower, 1975, p. 437; or Taylor, 1960). In contrast, other psychologists attach somewhat different meanings. The psychological literature sometimes employs the term "algorithmic" to mean "straightforward" and "heuristic" to mean "open-ended" or "unobvious" (see, e.g., Amabile, 1983a; or McGraw, 1978). In this book, the terminology will be consistent with the computer scientists'. In particular we will continue to use the terms "easy" and "difficult" to denote respectively what some psychologists refer to as "algorithmic" and "heuristic."

P. 12   "A Study of 58 major innovations . . .": Jewkes et al. (1969), pp. 96–98.
   "Ignorance is power . . .": Jewkes et al. (1969), p. 96.
   ". . . most of the engineers were individuals . . .": Jewkes et al. (1969), pp. 262–266.

P. 13   Quote: "Vertical thinking digs the same hole deeper . . .": de Bono (1968), pp. 4–5.

P. 14   ". . . advances in science occur through a series . . .": Kuhn (1962), sec. I.
   Quote by Simon: Simon (1977), p. 302.

P. 15   Quote by Hofstadter: Hofstadter (1985), p. 249.
   ". . . no idea is 100 percent novel": Guilford (1977), p. 108.
   ". . . innovative solutions usually evolve incrementally . . .": Weisberg (1986), p. 11.

Chapter 3

P. 19 "... the brain seems to require roughly 10 seconds ...": Simon (1969), pp. 78–79.

P. 22 "... group of five engineers ... was established ...": Kidder (1981), pp. 35–46.

"... the machine accounted for perhaps 10 percent ...": Kidder (1981), pp. 269–289.

P. 23 Quote: "More software projects have gone awry for lack of calendar time ...": Brooks (1975), p. 14.

P. 24 Quote: "Certain prolific persons are responsible for a disproportionate share of the achievements ...": Simonton (1984), p. 181.

Chapter 4

P. 25 Quote by Gleick: Gleick (1987), p. 175.

P. 26 "... birds have functional requirements other than airborne motion ...": Suh (1990), Ch. 7.

"... quality of solutions tends to increase ...": Weisberg (1986), p. 66.

"The objectives form a moving target ...": Simon (1977), pp. 239–240.

P. 28 Quote by Cox: Cox (1926), p. 218.

"A major tool to promote creativity ...": Adams (1979), p. 143.

P. 30 "Mathematical and verbal representations together form a more powerful tool ...": Adams (1979), p. 88.

Definition of *bisociation*: Koestler (1964).

"... solve a simpler problem when the original task appears too formidable ...": Polya (1945), p. 103.

P. 31 "For 13 days in October 1962 ...": Allison (1971).

P. 32 "The past served to guide present action for future purpose": Neustadt and May (1986), p. 41.

"The history of science suggests a series of stages ...": Kim (1990), Ch. 1.

Quote by Poincaré: Poincaré (1946), p. 385. Italics in original.

P. 33 "Poincaré's 'sensual imagery' led him to sense a mathematical proof": A. I. Miller (1984), p. 221.

P. 34 Quote by Einstein: Albert Einstein in a letter to Hadamard. In Hadamard (1945), pp. 142–143.

P. 35 "... drawing can be an integral part of the idea-generation process": Hanks and Belliston (1977), p. 6.

"The utility of externalization lies in the fact that ...": Stock (1985), p. 90.

Reasons for the utility of externalization: Wickelgren (1974), pp. 186–187.

Chapter 5

P. 37 Quote by Le Guin: Le Guin (1974), p. 100.

Steps involved in resolving a difficult task: Harrisberger (1966); Poincaré (1946), pp. 389–390; Wallas (1926), Ch. 4.

Incubation as conscious or subconscious activity: Some would argue that incubation is a superfluous concept since it implies the workings of unconscious mechanisms (Weisberg, 1986). In this section we use the term as a synonym for intermission and thereby side-step the issue of subconsciousness, a realm that is only poorly understood.

P. 39 **"Some referent disciplines to investigate relate to materials science . . .":** Kim (1988).

Quote: "Almost always the men who achieve these fundamental inventions . . .": Kuhn (1962), p. 90.

P. 40 **"77 percent of innovations spring from users . . .":** von Hippel (1985), p. 4. In a number of other fields, the bulk of the innovations are due to the manufacturers or suppliers. A case in point is engineering plastics, wherein 90% of the innovations result from the manufacturer and only 10% from users.

"Chester Carlson . . . invented a method to copy documents . . .": Jacobson and Hillkirk (1986), pp. 5, 198.

". . . Carlson attempted to market his idea . . .": Jacobson and Hillkirk (1986), p. 54.

P. 41 **"This approach relies on a number of questions . . .":** Osborn (1963).

"The morphological approach . . . has been used to explore the varieties . . .": Zwicky (1969), pp. 191–196.

P. 42 **Quote: "In all his explorations the morphologist strives for complete field coverage . . .":** Zwicky (1969), p. 44.

P. 43 **". . . brainstorming is based on the simple assumption that quantity can lead to quality . . .":** Osborn (1963), p. 146.

P. 44 **". . . quality of ideas can be higher *without* brainstorming . . . ";** Stein, v. 2 (1975), p. 61.

". . . solutions obtained from individuals working alone is often superior . . ."; Weisberg (1986), pp. 62–68.

Principles of synetics: Gordon (1961), Ch. 2.

P. 46 **"The value of group interaction . . . is well known":** Gordon (1961), Ch. 1.

P. 47 **"Picasso pursued many different avenues . . .":** Weisberg (1986), p. 125.

Steinberg's four strategies for metaplanning: Sternberg (1986), pp. 73–74.

Chapter 6

P. 51 **Quote by Watson:** Watson (1968), p. ix.

Quote: "However romantic and heroic we find the moment of discovery, we cannot believe . . .": Langley *et al.* (1987), pp. 3–4.

P. 52 **"Research is a more chaotic, jumbled exercise . . .":** Stock (1985), p. 7.

P. 53 **Quote by Polya:** Polya (1945), p. 106.

Quote by Brooks: Brooks (1975), p. 116. Italics in original.

"Problem solving is a process of selective trial and error": Simon (1977), p. 302.

Quote: "Personality factors are extremely important determiners of achievement . . .": Terman and Oden (1959), pp. 148–149.

P. 54 **Quote: "Perhaps the odds that any single contribution will prove successful are constant . . .":** Simonton (1984), p. 83.

P. 55 **Quote: *"The man-month as a unit for measuring the size of a job is a dangerous and deceptive myth . . .":*** Brooks (1975), p. 16. Italics in original.

P. 59 **The breadth-versus-depth trade-off occurs not only in searching for new knowledge . . .":** Guilford (1977), p. 126.

P. 60 **Quote: "To check out one idea at a time has several disadvantages . . .":** Pinchot (1985), p. 101.

P. 61 **Thoreau's philosophy of simplicity:** "Our life is frittered away by detail . . . . Simplicity! I say, let your affairs be as two or three, not a hundred or a thousand." Walden (1854), as given in Thoreau (1939), p. 97.

Chapter 7

P. 63 **Quote by Wiener:** Wiener (1967), p. 257.

P. 65 **"The choice of a worthwhile topic is a critical aspect of the research endeavor . . .":** "Intellectual competence must entail a set of skills of problem solving . . .[and] the potential for finding or creating problems—thereby laying the groundwork for the acquisition of new knowledge." Gardner (1983), pp. 60–61; italics in original.

Quote by Hadamard: Hadamard (1945), p. 126.

P. 67 **"The thesis often discusses the paths that led to nowhere . . .":** Stock (1985), p. 136.

"Do not wait for the grand, perfect idea to streak into your consciousness; it will never come . . .": "Successful entrepreneurs do not look for the 'biggie,' the idea that will . . . 'make one rich overnight.'" Drucker (1985), p. 34.

P. 68 **"Conducting research and writing about it are synergistic activities":** Stock (1985), p. 146.

". . . research can be an exhilarating experience": The joys of creative problem solving are manifold, not only in the product but in the process. In the words of a creativity consultant, "I like to think of creative thinking as the 'sex of our mental lives.' Ideas, like organisms, have a cycle. They are born, they develop, they reach maturity, and they die. So we need a way to generate new ideas. Creative thinking is that means, and like its biological counterpart, it is also pleasurable." von Oech (1983), p. 5.

While this is an interesting metaphor, it deserves some qualification. More precisely, ideas do not die but their utility may vanish. Some ideas may grow old and decrepit, while others remain forever young and fresh. The concept of a slide rule may wither under the weight of the electronic calculator; the principles of arithmetic, on the other hand, are likely to endure indefinitely.

Quote by Einstein: Einstein (1934), p. 2.

Chapter 8

P. 70 **Quote by Gardner:** Gardner (1982), p. 355.

"The creative person must have a healthy dose of confidence . . .": Torrance (1962), p. 74.

P. 71 **"Studies . . . indicate the dubious impact of education on eminence":** Cox (1926); Simonton (1984), pp. 44, 65.

Relationship between eminence and education: One explanation for the concave curve is that doctorate holders represent only a small proportion of the population today, and a minuscule fraction of yesteryear's. Given the chancy nature of creative thinking, the larger population having limited schooling may well generate more geniuses in absolute numbers, even if the proportion of achievers is actually less than that of the academically exalted group. In other words, higher education may still be conducive to promoting creativity, or at least not detrimental.

This argument, although it might be valid, is not entirely convincing or comforting. Advanced education represents a great investment of effort both for the student and the society at large. Should it not serve, among other things, as a vehicle for promoting innovative thinking? And should it not serve this goal so well as to eradicate all doubts about its efficacy? These goals are challenging, but ones that richly deserve to be addressed in a coherent fashion.

Quote by Cox: Cox (1926), p. 219.

P. 72   **Quote: "Time spent in pursuit of academic honors is time lost for the acquisition of information . . . not related to schoolwork . . .":** Simonton (1984), p. 74.

P. 73   **Quote: "Some companies give employees the freedom to use a percentage of their time on projects of their own choosing . . .":** Pinchot (1985), p. 198.

Quote: "True 'freedom' is not the absence of structure . . .": Kantner (1983), p. 248.

Necessity of balancing laissez-faire policy with structure: A judicious mix of intervention and free rein seems appropriate for diverse domains. For example, psychological research supports the popular notion that guidance and independence are both necessary for fully developing the artistic potential in children. (Gardner, 1982, p. 217.)

P. 74   **Two conditions for creative work:** Amabile (1988), pp. 144–145.

". . . open-ended tasks require intrinsic motivation": Hennesey and Amabile (1987).

". . . the supervisor must shield the supervisee from administrivial distractions . . .": Kantner (1983), p. 229.

P. 77   **"The unknown 'X' factor . . . is a regular participant in scientific discovery . . .":** Jewkes et al. (1969), p. 169; Stock (1985), p. 93.

P. 78   **"Alexander Fleming was conducting experiments on a virulent strain of bacteria . . .":** Macfarlane (1984), pp. 118–120.

"The new agent—penicillin—came from . . . a variant so rare . . .": Macfarlane (1984), pp. 134, 138.

P. 82   **"The supervision of creative talent requires respect for the supervisee . . .":** Torrance (1962), p. 170.

P. 83   **"If a subject finds intrinsic interest in the task *and* the problem is difficult . . .":** Amabile (1983), p. 15; McGraw (1978).

Chapter 9

P. 85   **Quote by Thoreau:** Thoreau (1849), "Wednesday"; as given in Thoreau (1893), p. 385.

Appendix A

P. 88   **Quote by Poincaré:** *Valeur de la Science* (1904), as given in Poincaré (1946), p. 208.

P. 90   **". . . a paradigmatic result refers to a shift in our views":** Kuhn (1962), Ch. 1.

P. 91   **Applicability of the Central Limit Theorem:** The mathematically inclined reader may be aware that according to the Central Limit Theorem, the Gaussian distribution is pervasive in the following sense. Given any set of random variables that are mutually independent and each of which has a finite variance, their sum tends toward a Gaussian distribution when the set is large. This result is unchanged by the fact that the independent variables may have similar or dissimilar probability distributions, any (or none) of which might be Gaussian.

". . . amino acids . . . can be formed from a broth of lifeless chemicals when exposed to a flux of energy": Miller (1957).

P. 92   "An important class of results relates to interdimensional trade-offs": Kim (1990), Ch. 2.

Appendix B

P. 93   Quote by White: Byrne (1988), p. 383.

"A study . . . cites humor and playfulness as distinguishing characteristics of creative individuals . . .": Torrance (1962), p. 122.

Definition of humor: *Webster's Ninth New Collegiate* (1984).

P. 94   ". . . laughter results primarily from an enjoyment of observing the weaknesses or misfortunes of others": *Republic*, Bk. X, sec. 606; as given in Plato (1928), p. 405.

". . . recognized ridicule and the ability to laugh at oneself as other sources of comedy": *Symposium*, sec. 193; in Plato (1947), p. 147.

"Aristotle . . . confirmed the abusive character of jokes": *Nicomachean Ethics*, Bk. IV, sec. 8; in Aristotle (1980), p. 103.

"Surprise may be due to some disappointed expectation . . .": Aristotle, *Rhetoric*, Bk. III, sec. xi; as given in Barnes (1984), p. 2253.

". . . laughter springs from an innocuous type of ugliness": Aristotle, *Poetics*, sec. v; as given in Barnes (1984), p. 2319.

". . . risibility springs primarily from unseemly things . . .": *De Oratore*, Bk. II, secs. lvii–lxi; as given in Cicero (1942), pp. 373–383.

". . . laughter can only result from . . . hatred or surprise . . .": Descartes, *Passions de L'Ame* (1649), Pt. 2, art. CXXVI and Pt. 3, art. CLXXVIII; as given in Haldane and Ross (1955), pp. 386 and 413.

". . . laughter springs from self-satisfaction . . .": *Leviathan*, Part I, Ch. 6; in Hobbes (1904), p. 34.

". . . people can bear to be criticized outright, but not mocked or ridiculed": Preface to *Tartuffe* in Molière (1965), p. 12.

"Locke . . . defined wit as a collection of ideas whose mutual resemblance yields pleasant images . . .": Locke, *An Essay Concerning Human Understanding* (1960 and later editions), Ch. XI; as given in Locke (1975), p. 156.

Quote by Johnson: "every dramatic composition which raises mirth, is comick": Johnson (1824), p. 217.

P. 95   ". . . laughter results from joy rather than contempt or vanity": *Dictionniare Philosophique*, given in Voltaire (1901), pp. 58–59.

"Kant . . . identified laughter with the abrupt change in physiological processes . . .": *Critique of Judgement*, Pt. I, sec. 54; given in Kant (1914), pp. 221–223.

". . . our species is the only one capable of joking . . .": Emerson (1876), p. 139.

". . . the comic . . . relates to something human . . .": Bergson (1911), p. 3.

"A necessary condition for humor is indifference . . .": Bergson (1911), p. 4.

". . . laughter is a social phenomenon that calls for a feeling of kinship . . .": Bergson (1911), p. 6.

"A common source of laughter is the discrepancy between human intention and the state of the world": Bergson (1911), pp. 10, 21.

". . . laughter results from inflexible responses to situations that call for versatility . . .": Bergson (1911), p. 198.

Freud's systematic theory of levity: Freud (1960), p. 236.

Freud's identification of two types of jokes: Freud (1960), pp. 90; 100.

Quote: "Time is nature's way of keeping everything from happening at once": Byrne (1988), p. 98.

Quote: "If I had known I was going to live this long . . .": Byrne (1988), p. 100.

Quote by Kissinger: "University politics are vicious . . .": Kissinger, as given in Byrne (1988), p. 159.

"A common source of comedy is a sense of superiority . . .": Freud (1960), pp. 199–200.

P. 96 ". . . a joke is a comic event that originates from the unconscious": Freud (1960), p. 208.

". . . an observer may find comic pleasure in solitude . . .": Freud (1960), p. 143.

"Bypassing the censoring mechanisms of rationality . . . allows for the genesis of new ideas . . .": Freud (1960), Ch. VI.

". . . humor occurs when an observer is led to develop a sense of discomfiture . . .": Freud (1960), p. 229.

"Freud regarded humor as the highest of the defense mechanisms": Freud (1960), p. 233.

Eastman's principles for jokes: Eastman (1921), pp. 88–116.

P. 97 Quote by Twain: "Always do right . . .": Twain, in an address to the Young People's Society, Brooklyn, 16 February 1901; as given in Beck (1980), p. 626.

Quote by Shakespeare: Shakespeare, *Hamlet,* II, 2.

Quote: "Victory finds a hundred fathers . . .": Count Galeazzo Ciano in *The Ciano Diaries* (1946); as given in Beck (1980), p. 857.

Quote by Woody Allen: "More than any time in history mankind faces a crossroads . . .": Allen, as given in Byrne (1988), p. 82.

P. 98 Quote by Austen: "What dreadful hot weather we have . . .": Austen, in a letter dated 1796; as given in Byrne (1988), p. 112.

Quote by Twain: "The reports of my death are greatly exaggerated": Twain, in a cable from London to the Associated Press, 1897; quoted in Beck (1980), p. 625.

Quote by Wilde: "Men always want to be a woman's first love . . .": *A Woman of No Importance,* Act II; as given in Wilde (1905), p. 29.

Quote by Twain: "Familiarity breeds contempt . . .": Twain (1935), p. 237.

Quote: "The difference between Los Angeles and yogurt . . .": Byrne (1988), p. 88.

Quote: "Husbands are like fires . . .": Zsa Zsa Gabor, as given in Byrne (1988), p. 86.

P. 99 Quote by Bok: "If you think education is expensive . . .": Bok, as given in Byrne (1988), p. 158.

Story of an emperor touring his provinces: Freud (1960), pp. 68–69.

Oscar Wilde anecdote: Redman (1959), p. 110.

P. 100 Freud's conditions of the comic situation: Freud (1960), pp. 218–221.

Example of "gallows humor": Freud (1960), p. 229.

Appendix C

P. 102 **Quote by Jung:** Jung (1971), pp. 220–221.

"... **artists often reach their peak of productive output earlier than scientists ...**": Dennis (1966).

Peaks of productive output for various fields: Lehman (1953), pp. 324–327.

Modal productive ages for various types of writers: Simonton (1975).

Average age at which various types of writing were produced: Lehman (1953), pp. 118, 120, 132.

P. 103 "... **in intellectual fields ... the age of greatest contribution has been decreasing ...**": Lehman (1953), p. 327.

"**National Academy of Sciences rose from 41.3 years ... to 51.8 ...**": Lehman (1953), p. 278.

"... **average age of political leaders has been increasing ...**": Lehman (1953), p. 327.

"... **median age of cabinet members has increased ...**": Lehman (1953), p. 272.

Two-component model of creativity's relation to age: Simonton (1983); Simonton (1984), p. 109.

Model depicting creativity as a function of age: The assumptions can be transformed into a set of linear differential equations. The solution is given by $p(t) = c[exp(-at) - exp(-bt)]$, where $t = Age - 20$, is the age of the individual shifted by 20 years, $p(t)$ is the number of tasks completed in year t, and exp is the exponential function. The other parameters define the specific shape of the curve: a is the number of tasks addressed per year, b specifies the rate of completion ($b > a$), and c defines the height of the curve to allow for productivity differences. The parameters a and b are deemed to depend on the domain of inquiry, and c on the individual. Typical values are $a = .04$, $b = .05$, and $c = 61$. The fit between the data and predicted curve has correlation values in the upper .90s (Simonton, 1982; Simonton, 1984, p. 110).

P. 105 "**The capacity of short-term memory is also very limited ...**": Miller (1956).

P. 108 **Quote by James:** James (1907), pp. 295–296.

Appendix D

P. 109 **Quote by Lucientes:** *La fantasia abandonada de la razon produce monstruos imposibles: unida con ella, es madre de las artes y origen de sus marabillas."* From *Los Caprichos* (1799), Plate 43; as given in Lopez-Rey (1970), p. 200.

"**Creativity springs from the deliberate selection of options ... :** "It is useful to view creating as a process of selecting from among the many possible outcomes— arrays of words, formulas, pigments on a surface, and so on. . . . The resources of mind—noticing, realizing, directed remembering, problem finding, schemata, hill climbing, critical reasons, and many more—each contribute to creating by helping to accomplish selection." Perkins (1981), pp. 286–287.

Definition of agent: *Webster's Ninth New Collegiate* (1987).

P. 110 "**Creativity Support System ... is tailored after a system developed for more limited objectives ...**": Kim *et al.* (1988).

Selected Bibliography

Adams, J.L. *The Care and Feeding of Ideas*. Reading, MA: Addison-Wesley, 1986.

Adams, J.L. *Conceptual Blockbusting*, 2nd ed. NY: Norton, 1979.

Allen, M.S. *Morphological Creativity*. Englewood Cliffs, NJ: Prentice-Hall, 1962.

Allison, Graham T. *Essence of Decision: Explaining the Cuban Missile Crisis*. Boston: Little, Brown, 1971.

Amabile, Teresa M. *The Social Psychology of Creativity*. NY: Springer-Verlag, 1983a.

Amabile, Teresa M. "The Social Psychology of Creativity: A Componential Conceptualization." *J. of Personality and Social Psychology*, v. 45(2), 1983b.

Amabile, Teresa M. "A Model of Creativity and Innovations in Organizations." *Research in Organizational Behavior*, v. 10, 1988, pp. 123–167.

Aristotle. *The Nichomean Ethics*. New York: Oxford Univ. Press, 1980.

Barnes, Jonathan. *The Complete Works of Aristotle*, v. 2. Princeton: Princeton Univ. Press, 1984.

Baron, J.B., and R.J. Sternberg, eds. *Teaching Thinking Skills: Theory and Practice*. NY: Freeman, 1987.

Beck, Emily Morison, ed. *Bartlett's Familiar Quotations*. Boston: Little, Brown, 1980.

Bergson, Henri. *Laughter*. Trans. by C. Brereton and F. Rothwell. London: Macmillan, 1911.

Bittinger, M. *Logic, Proof and Sets*. Reading, MA: Addison-Wesley, 1982.

Bohm, D., and F. D. Peat. *Science, Order, and Creativity*. NY: Bantam, 1987.

Brooks, F. P. Jr. *Mythical Man-Month*. Reading, MA: Addison-Wesley, 1975.

Burgelman, R.A., and M.A. Maidique. *Strategic Management of Technology and Innovation*. Homewood, IL: Irwin, 1988.

Byrne, Robert. *1,911 Best Things Anybody Ever Said*. NY: Ballantine, 1988.

Cattell, R. B., and J. Butcher. *The Prediction of Achievement and Creativity*. Indianapolis: Bobbs Merrill, 1968.

Cicero, Marcus Tullius. *De Oratore*, Books I and II. Trans. by E. W. Sutton. Cambridge, MA: Harvard Univ. Press, 1942.

Cox, C. M. *The Early Mental Traits of Three Hundred Geniuses*. Stanford, CA: Stanford Univ. Press, 1926.

Davidson, J.E., and R.J. Sternberg. "The Role of Insight in Intellectual Giftedness." *Gifted Child Quarterly*, v. 28, 1984, pp. 8–64.

de Bono, Edward. *New Think: The Use of Lateral Thinking in the Generation of New Ideas*. NY: Basic, 1968.

de Bono, Edward. *Lateral Thinking: Creativity Step by Step*. NY: Harper & Row, 1970.

Dennis, W. "Creative Productivity between the Ages of 20 and 80 Years." *Journal of Gerontology*, v. 21,1966, pp. 106–114.

Dickens, Charles. *The Pickwick Papers*. London: J.M. Dent & Sons, 1945.

Drucker, Peter F. *Innovation and Entrepreneurship*. NY: Harper & Row, 1985.

Eastman, Max. *The Sense of Humor*. NY: Scribner's, 1921.

Ebbinghaus, H. *Memory*. NY: Dover, 1885 and 1964.

Einstein, Albert. *Essays in Science*. Trans. by Alan Harris. NY: Philosophical Library, 1934.

Ellinger, J.H. *Design Sythesis*, v. 1. NY: John Wiley and Sons, 1968.

Emerson, Ralph Waldo. *Letters and Social Aims*. Boston: Osgood, 1876.

Feldhusen, John F., Donald J. Treffinger, and Susan J. Bahlke. "Developing Creative Thinking: The Purdue Creativity Program." *J. of Creative Behavior*, v. 4(2), 1970, pp. 85–90.

Forsyth, R., and C. Naylor. *The Hitch-Hiker's Guide to Artificial Intelligence. NY: Chapman and Hall, 1985.*

Freud, Sigmund. *Jokes and their Relation to the Unconscious*. Trans. by J. Strachey. NY: Norton, 1960.

Galton, Francis. *Hereditary Genius*, 2nd ed. London: Macmillan, 1892.

Gardner, Howard. *Art, Mind and Brain*. NY: Basic, 1982.

Gardner, Howard. *Frames of Mind*. NY: Basic, 1983.

Ghiselin, B. *The Creative Process*. Berkeley: University of California Press, 1952.

Glegg, G. L. *The Science of Design*. London: Cambridge Univ. Press, 1973.

Gleick, James. *Chaos: Making a New Science*. NY: Viking, 1987.

Glorioso, R. M., and F. C. C. Osorio. *Engineering Intelligent Systems, Concepts, Theory and Applications*. Bedford, MA: Digital Press, 1980.

Gordon, W. J. J. *Synectics*. NY: Harper & Row , 1961.

Grossman, Stephen R., Bruce E. Rodgers, and Beverly R. Moore. *Innovation, Inc.: Unlocking Creativity in the Workplace*. Plano, TX: Wordware, 1988.

Guilford, J.P. *Way Beyond the IQ: Guide to Improving Intelligence and Creativity*. Buffalo, NY: Creative Education Foundation, 1977.

Hadamard, Jacques. *An Essay on the Psychology of Invention in the Mathematical Field*. NY, Dover, and Princeton: Princeton Univ. Press, 1945.

Haldane, Elizabeth S., and G.R.T. Ross. *The Philosophical Works of Descartes*, v. I. NY: Dover, 1955.

Hanks, Kurt, and Larry Belliston. *Draw! A Visual Approach to Thinking, Learning, and Communicating*. Los Altos, CA: William Kaufmann, 1977.

Harrisberger, L. *Engineersmanship*. Belmont, CA: Brooks/Cole Publishing, 1966.

Hayes, John R. *The Complete Problem Solver*. Philadelphia: Franklin Institute, 1981.

Hennessey, Beth A. "The Effect of Extrinsic Constraints on Children's Creativity While Using a Computer." Working Paper, Dept. of Psychology, Wellesley College, Wellesley, MA: July 1988.

Hennessey, Beth A., and Teresa M. Amabile. *Creativity and Learning*. Washington, DC: National Education Assoc., 1987.

Hilgard, E.R., and G.H. Bower. *Theories of Learning*, 4th ed. Englewood Cliffs, NJ: Prentice-Hall, 1975.

Hobbes, Thomas. *Leviathan*. Cambridge: Cambridge Univ. Press, 1904.

Hofstadter, D.R. *Metamagical Themas*. NY: Basic, 1985.

Jacobsen, Gary, and John Hillkirk. *Xerox: American Samurai*. New York: Macmillan, 1986.

James, W. *Psychology*. NY: Henry Holt and Co., 1907.

Jenkins, John G., and Karl M. Dallenbach. "Obliviscence During Sleep and Waking." *American J. of Psychology*, v. 35, 1924, pp. 605–612.

Jewkes, John, David Sawers, and Richard Stillerman. *The Sources of Invention*, 2nd ed. NY: Macmillan, 1969.

Johnson, Samuel. *The Rambler*. London: Jones, 1824.

Jung, Carl Gustav. *Psychological Types: The Collected Works of Jung, v. 6*. Ed. by H. Read, M. Fordham, G. Adler, and W. McGuire. Princeton: Princeton Univ. Press, 1971.

Kant, Immanuel. *Critique of Judgement*. Trans. by J.H. Bernard. London: Macmillan, 1914.

Kanter, R.M. *The Change Masters*. NY: Simon and Schuster, 1983.

Keeney, R. L., and H. Raiffa. *Decisions with Multiple Objectives: Preferences and Value Trade-offs*. NY: John Wiley and Sons, 1976.

Keil, John M. *The Creative Mystique*. (Cassette.) NY: Wiley, 1986.

Kidder, Tracy. *The Soul of a New Machine*. Boston: Little, Brown, 1981.

Kim, S.H. "Mathematical Foundations of Manufacturing Science: Theory and Implications." Ph.D. thesis, M.I.T., May 1985.

Kim, S.H. "A Unified Architecture for Design and Manufacturing Integration," Technical Report, Lab. for Manufacturing and Productivity, M.I.T., Cambridge, MA,1987.

Kim, S.H. "Research in Manufacturing Systems: Framework and Referent Paradigms." Technical Report, Laboratory for Manufacturing and Productivity, M.I.T., Cambridge, MA, 1988.

Kim, S.H., "Difficult Problems and Creative Solutions." *International J. of Computer Applications in Technology*, v. 2(3), 1989, pp. 171–185.

Kim, S.H. *Designing Intelligence: A Framework for Smart Systems*. NY: Oxford Univ. Press, 1990.

Kim, S.H., and N.P. Suh. "Mathematical Foundations for Manufacturing," Trans. ASME/*J. of Engineering for Industry,* v. 109(3), 1987, pp. 213–218.

Kim, S.H., S. Hom, and S. Parthasarathy. "Design and Manufacturing Advisor for Turbine Disks." *Robotics and Computer-Integrated Manufacturing*, v. 4(3/4), 1988, pp. 585–592.

Kneebone, G.T. *Mathematical Logic and the Foundations of Mathematics*. NY: van Nostrand, 1963.

Kneller, G. F. *The Art and Science of Creativity*. NY: Holt, Rinehart and Winston, 1965.

Koestler, A. *The Act of Creation*. NY: Macmillan, 1964.

Kosslyn, S.M. *Image and Mind*. Cambridge, MA: Harvard Univ. Press, 1980.

Kowalski, R. *Logic for Problem Solving*. NY: North Holland, 1979.

Kuhn, T. S. *The Structure of Scientific Revolutions*, 2nd ed. Chicago: University of Chicago Press, 1962, 1970.

Langley, P., H.A. Simon, G.L. Bradshaw, and J.M. Zytkow. *Scientific Discovery: Computational Explorations of the Creative Process*. Cambridge, MA: MIT Press, 1987.

Le Guin, U. K. *The Dispossessed*. NY: Harper & Row, 1974.

Lehman, H. C. *Age and Achievement*. Princeton: Princeton Univ. Press, 1953.

Lenat, D. "AM: Discovery in Mathematics as Heuristic Search." In R. Davis and D.B. Lenat, eds., *Knowledge-Based Systems in Artificial Intelligence*. NY: McGraw-Hill, 1982 (Based on Ph.D. thesis, Stanford University, CA, 1977).

Lenat, D. "EURISKO: A Program That Learns New Heuristics and Design Concepts: The Nature of Heuristics, III: Program Design and Results." *Artificial Intelligence*, v. 21 (2), 1983: pp. 61–98.

Locke, John. *An Essay Concerning Human Understanding*. Ed. by P.H. Nidditch. Oxford: Oxford Univ. Press, 1975.

Lopez-Rey, Jose. *Goya's Caprichos*, v.1. Westport, CT: Greenwood, 1970.

Macfarlane, Gwyn. *Alexander Fleming, The Man and the Myth*. Cambridge, MA: Harvard Univ. Press, 1984.

Madigan, Carol Orsag, and Ann Elwood. *Brainstorms and Thunderbolts*. NY: Macmillan, 1983.

McGraw, Kenneth O. "The Detrimental Effects of Reward on Performance: A Literature Review and Prediction Model." In M.R. Lepper & David Greene, eds., *The Hidden Costs of Reward*. Hillsdale, NJ: Lawrence Erlbaum Associates, 1978, pp. 33–60.

McKim, Robert H. *Experiences in Visual Thinking*, 2nd ed. Belmont, CA: Wadsworth, 1980.

Miller, A.I. *Imagery in Scientific Thought*. Cambridge, MA: Birkhauser, 1984.

Miller, G.A. "The Magical Number Seven, Plus or Minus Two: Some Limits on our Capacity for Processing Information." *Psychological Review*, v. 63, 1956, pp. 81–97.

Miller, Stanley L. "The Mechanism of Synthesis of Amino Acids by Electrical Discharges." *Biochimica et Biophysica Acta*, v. 23(3), 1957, pp. 480–489.

Minsky, M.L. *The Society of Mind*. NY: Simon & Schuster, 1986.

Molière. *Tartuffe*. Trans. by R.W. Hartle. Indianapolis: Bobbs-Merrill, 1965.

Moskowitz, Robert A. *Creative Problem Solving*. (Workbook and cassettes.) NY: American Management Assoc., 1978.

Neustadt, Richard E., and Ernest R. May. *Thinking in Time: The Uses of History for Decision Makers*. NY: Free Press, 1986.

Newell, A., and H. A. Simon. *Human Problem Solving*. Englewood Cliffs, NJ: Prentice-Hall, 1972.

Ogilvy, David. *Confessions of an Advertising Man*. (Cassettes.) Englewood, CO: Newstrack, 1984.

Osborn, A. F. *Applied Imagination*, 3rd ed. NY: Charles Scribner & Sons, 1963.

Parnes, S.J., and H.F. Harding. *A Source Book for Creative Thinking*. NY: Charles Scribner & Son, 1962.

Perkins, D.N. *The Mind's Best Work*. Cambridge, MA: Harvard Univ. Press, 1981.

Pinchot, Gifford, III. *Intrapreneuring*. NY: Harper & Row, 1985.

Plato. *The Republic*. NY: Charles Scribner & Son, 1928.

Plato. *Plato in Twelve Volumes, v. III: Lysis, Symposium, Gorgias*. Trans. by W.R.M. Lamb. Cambridge, MA: Harvard Univ. Press, 1947.

Poincaré, Henri. *The Foundations of Science: Science and Hypothesis; The Value of Science; Science and Method*. Trans. by G.B. Halsted. Lancaster, PA: Science Press, 1946.

Polya, G. *How to Solve It*. Princeton, NJ: Princeton Univ. Press, 1945.

Redman, Alvin, ed. *The Wit and Humor of Oscar Wilde*. NY: Dover, 1959.

Rosen, J. *A Symmetry Primer for Scientists*. NY: Wiley, 1983.

Rowan, R. *The Intuitive Manager*. NY: Berkley, 1986.

Schlesinger, Arthur Jr. *Creativity in Statecraft*. Washington, D.C.: Library of Congress, 1983.

Sculley, John. *Odyssey*. NY: Harper & Row. 1987.

Shepard, Roger N., and Lynn A. Cooper. *Mental Images and Their Transformations*. Cambridge, MA: MIT Press, 1982.

Simon, Herbert Alexander. *Models of Discovery*. Boston: Reidel, 1977.

Simon, Herbert Alexander. *The Sciences of the Artifical*. Cambridge, MA: M.I.T. Press, 1969, 1981.

Simonton, Dean Keith. "Age and Literary Creativity: A Cross-Cultural and Transhistorical Survey." *Journal of Cross-Cultural Psychology*, v.6 (3), 1975, pp. 259–277.

Simonton, Dean Keith. "Creative Productivity and Age: A Mathematical Model Based on a Two-Step Cognitive Process." *Developmental Review*, v. 3, 1983, pp. 77–111.

Simonton, Dean Keith. *Genius, Creativity and Leadership*. Cambridge, MA: Harvard Univ. Press, 1984.

Soloway, E., and K. Ehrlich. "Empirical Studies of Programming Knowledge." *IEEE Trans. Software Engineering*, SE-10(5), 1984, pp. 595–609.

Stein, M.I. *Stimulating Creativity, v. 1: Individual Procedures; v.2: Group Procedures.* NY: Academic Press, 1975.

Sternberg, R.J. "Criteria for Intellectual Skills Training." *Educational Researcher,* v.12, 1983, pp. 6–12.

Sternberg, R.J. *Intelligence Applied: Understanding and Increasing your Intellectual Skills.* San Diego: Harcourt Brace Jovanovich, 1986.

Sternberg, R.J. *Nature of Creativity: Contemporary Psychological Perspectives.* Cambridge: Cambridge Univ. Press, 1988.

Stock, Molly W. *A Practical Guide to Graduate Research.* NY: McGraw-Hill, 1985.

Suh, Nam P. *The Principles of Design.* New York: Oxford Univ. Press, 1990.

Suh, N. P., A. C. Bell, and D. C. Gossard. "On An Axiomatic Approach to Manufacturing and Manufacturing Systems." *J. of Engineering for Industry,* Trans. ASME, v. 100(2), May 1978: pp.127–130.

Taylor, D. W. "Toward an Information Processing Theory of Motivation." In M. R. Jones (Ed.), *Nebraska Symposium on Motivation.* Lincoln: University of Nebraska Press, 1960, pp. 13–79.

Terman, L. M. *Genetic Studies of Genius. v. I: Mental and Physical Traits of a Thousand Gifted Children.* Stanford: Stanford Univ. Press, 1925.

Terman, L. M., and M. H. Oden. *Genetic Studies of Genius. v. IV: The Gifted Child Grows Up,* 1947. *v. V: The Gifted Group at Mid-Life.* Stanford: Stanford Univ. Press, 1959.

Thoreau, Henry David. *A Week on the Concord and Merrimack Rivers.* Boston: Houghton-Mifflin, 1893.

Thoreau, Henry David. *Walden.* NY: Heritage Club, 1939.

Torrance, E. Paul. *Guiding Creative Talent.* Englewood Cliffs, NJ: Prentice-Hall, 1962.

Twain, Mark. *Mark Twain's Notebook.* NY: Harper, 1935.

Voltaire. *A Philosophical Dictionary: The Works of Voltaire, vol. VI, Part I.* Trans. by W. F. Fleming. NY: St. Hubert Guild, 1901.

von Hippel, E. *The Sources of Innovation.* Oxford: Oxford Univ. Press, 1985.

von Oech, Roger. *A Whack on the Side of the Head: How to Unlock your Mind for Innovation.* NY: Warner, 1983.

Wallas, Graham. *The Art of Thought.* NY: Harcourt Brace, 1926.

Walters, J.M., and Gardner, H. "The Theory of Multiple Intelligences: Some Issues and Answers." *Practical Intelligence,* R.J Sternberg and R.K. Wagner (eds). NY: Cambridge Univ. Press, 1986, pp.163–182.

Watson, J. D. *The Double Helix.* NY: New American Library, 1968.

Weisberg, Robert W. *Creativity: Genius and Other Myths.* NY: Freeman, 1986.

Weyl, H. *Symmetry.* Princeton: Princeton Univ. Press, 1952.

Wickelgren, Wayne A. *How to Solve Problems.* San Francisco: Freeman, 1974.

Wiener, Norbert. *The Human Use of Human Beings: Cybernetics and Society.* NY: Avon, 1967.

Wilde, Oscar. *The Plays of Oscar Wilde, v. 1.* Boston: Luce, 1905.

Zwicky, Fritz. *Discovery, Invention, Research through the Morphological Approach.* NY: Macmillan, 1969.

Index